Reasonable Books = Reasonable Prices

Irina Kleyman

AF875446

Exceptional for Home Schooling

Math Help
Grade 1

First Edition

Reasonable Books = Reasonable Prices

Irina Kleyman

Exceptional for Home Schooling

Math Help
Grade 1

I dedicate this book to my children and grandchildren.

Acknowledgments
Special thanks to Michael G LaFratta, Vladimir Kleyman, who has assisted in bringing all 10 books to market.

Author's note: In the many years that I've taught mathematics, I have been informed by many of my students that my methods for teaching have helped them tremendously in learning math. I have written these books in order to continue to help others learn math effectively.

Math Help Grade 1
Math Help Grade 2
Math Help Grade 3
Math Help Grade 4
Math Help Grade 5
Math Help Grade 6
Math Help Grade 7
Math Help Grade 8
Math Help for College (Summary Grades 1 – 5)
Math Help for University (Basic, Algebra, Geometry)

ISBN: 978-1-7349699-0-0 Paperback

Syllabus Math 1, 2, 3, 4

Grade 1 # 1 – 4   Grade 2 # 5 – 8   Grade 3 # 9-12   Grade 4 # 13 - 16

Syllabus includes topics, what students need to know.
Unknowns are numbers, which we need to find.

#	Grades	Topics. Advanced Problems (A), Pages (p)
	Chapters	Basic Problems (B), Chapter (C), Lesson (L)
1	1-1	One-Digit Numbers (B)
2	1-2	Two-Digit Numbers (B)
3	1-3	Three numbers (B)
4	1-4	Finding unknowns (A-1)
	Grade 1	Final Exam 1 B, C 4, L 12 # 1-20 odd, p 94
5	2-1	Addition or Subtraction Numbers (20–29) (B)
6	2-2	Addition or Subtraction Numbers (50–59) (B)
7	2-3	Addition or Subtraction Numbers (30–39) (B)
8	2-4	Finding unknowns (A-2)
	Grade 2	Final Exam 2 B, C 4, L 17 # 1-20 odd, p 106
9	3-1	Addition or Subtraction Numbers (60–69) (B)
10	3-2	Addition or Subtraction Numbers (90–99) (B)
11	3-3	Addition or Subtraction Numbers (70–79) (B)
12	3-4	Finding unknowns (A-3)
	Grade 3	Final Exam 3 B, C 4, L 19 # 1-20 odd, p 121
13	4-1	Three-Digit numbers (B)
14	4-2	Multiplication numbers (B)
15	4-3	Division numbers (B)
16	4-4	Finding unknowns (A-4)
	Grade 4	Final Exam 4 B, C 4, L 12 # 1 – 10 all, p 133

Parents, Tutors, Students Middle School, High School, Colleges, University can help primary school students. Please read slowly math books to primary school students.

Grade 1

www.mathbooksbyirina.com

Chapter 1

One – Digit

Numbers

Grade 1 Chapter 1 Contents
One-Digit Numbers
Lesson 1 – 8
Basic Problems
Buy in the learning store counting sticks for calculations (100).

1)Work with lesson 1, read rules and examples, 2) Solve the problems from contents, 3) Check answers in the end of chapter 1, 4) Work with the next lessons 2, 3, 4, 5, …, 5) Solve Test 1 A, 6) Grading Scale: A =10 points, B=9points, C=8 points, D=7 points, F=0-6 points, 7) If you solve 10 problems correctly, you get an A, 8) If have an A, you can move on to next chapter, 9) If your grade is lower than an A, work with the mistakes, then solve Test 1B, 10) When you finish all of the chapters, solve Final Exam in last chapter.

Grade 1 Chapter 1 Lesson 1
One-Digit Numbers
Basic Problems
Lesson 1
Numbers from 0 to 9

Count forward from 1 to 9 1, 2, 3, 4, 5, 6, 7, 8, 9
One, two, three, four, five, six, seven, eight, nine
First, second, third, fourth, fifth, sixth, seventh, eighth, ninth
Count backwards from 9 to 1 9, 8, 7, 6, 5, 4, 3, 2, 1
The smallest number is 0 The biggest number is 9
Equal means $5 = 5$ Number zero is less than one $0 < 1$

Less than (<) means

$0 < 1$	$1 < 2$	$2 < 3$	$3 < 4$	$4 < 5$
$5 < 6$	$6 < 7$	$7 < 8$	$8 < 9$	$9 = 9$

More than (>) means

$9 > 8$	$8 > 7$	$7 > 6$	$6 > 5$	$5 > 4$
$4 > 3$	$3 > 2$	$2 > 1$	$1 > 0$	$0 = 0$

To compare numbers find what number greater or smaller or numbers equal

0 ? 3 $0 < 3$ 3 ? 0 $3 > 0$

Put symbols <, > or = instead question mark

	a	b	c	d	e
# 1	0 ? 5	1 ? 5	2 ? 5	3 ? 5	4 ? 5
# 2	5 ? 5	6 ? 5	7 ? 5	8 ? 5	9 ? 5
# 3	0 ? 6	1 ? 6	2 ? 6	3 ? 6	4 ? 6
# 4	5 ? 6	6 ? 6	7 ? 6	8 ? 6	9 ? 6
# 5	0 ? 7	1 ? 7	2 ? 7	3 ? 7	4 ? 7
# 6	5 ? 7	6 ? 7	7 ? 7	8 ? 7	9 ? 7
# 7	0 ? 8	1 ? 8	2 ? 8	3 ? 8	4 ? 8
# 8	5 ? 8	6 ? 8	7 ? 8	8 ? 8	9 ? 8
# 9	0 ? 9	1 ? 9	2 ? 9	3 ? 9	4 ? 9
# 10	5 ? 9	6 ? 9	7 ? 9	8 ? 9	9 ? 9

Grade 1 Chapter 1 Lesson 2

Addition numbers in row

The first operation with numbers is addition +
To add numbers put numbers together: 2 + 3 = 5
How many in all or how many together? If put together 2 and 3 results is 5
If you need to add 5 + 3 you have to count after 5 forward 3 times
That should be 6, 7, 8, that is why 5 + 3 = 8
The sum of a number and zero is the same as the number: 5 + 0 = 5 0 + 5 = 5
Addition numbers in row 6 + 2 = 8, Count from 6 forward 2 times 7, 8

Add numbers Review 1 A

	a	b	c	d	e
# 1	0 + 0	1 + 0	2 + 0	3 + 0	4 + 0
# 2	5 + 0	6 + 0	7 + 0	8 + 0	9 + 0
# 3	0 + 1	1 + 1	2 + 1	3 + 1	4 + 1
# 4	5 + 1	6 + 1	7 + 1	8 + 1	1 + 8
# 5	0 + 2	1 + 2	2 + 2	3 + 2	4 + 2
# 6	5 + 2	6 + 2	7 + 2	2 + 7	1 + 7
# 7	0 + 3	1 + 3	2 + 3	3 + 3	4 + 3
# 8	5 + 3	6 + 3	0 + 5	1 + 5	2 + 5
# 9	0 + 4	1 + 4	2 + 4	3 + 4	4 + 4
# 10	5 + 4	3 + 5	4 + 5	0 + 6	2 + 6

Property of addition.

Two numbers can add in any order. The sum wills the same. 2+3=5 3+2=5
If you need to add 3 + 6, easy to start from bigger number, you need to start after 6 by count forward 3 times, that should be 7, 8, 9. That why 6 + 3 = 9
3 + 6 = 6 + 3 = 9

Add numbers. Review 1 B

	a	b	c	d	e
# 1	0 + 1	1 + 1	2 + 1	3 + 1	4 + 1
# 2	5 + 1	6 + 1	7 + 1	8 + 1	1 + 8
# 3	0 + 2	1 + 2	2 + 2	3 + 2	4 + 2
# 4	5 + 2	6 + 2	7 + 2	2 + 7	1 + 7
# 5	0 + 3	1 + 3	2 + 3	3 + 3	4 + 3
# 6	5 + 3	6 + 3	6 + 1	6 + 2	6 + 3
# 7	0 + 4	1 + 4	2 + 4	3 + 4	4 + 4
# 8	0 + 5	1 + 5	2 + 5	3 + 5	4 + 5
# 9	0 + 6	1 + 6	2 + 6	3 + 6	0 + 8
# 10	0 + 7	1 + 7	2 + 7	1 + 8	5 + 4

Grade 1 Chapter 1 Lesson 3
Subtraction numbers in row

Second operation with numbers is subtraction –
To subtract numbers, take away something: 5 – 2 = 3

How many are left? 3 are left

If you need to subtract 5 – 3 you have to count backward from 5 three times 4, 3, 2
This is why 5 – 3 = 2

Any number subtract zero result is the same number: 5 – 0 = 5
Any number subtract the same number result is zero: 5 – 5 = 0

Subtraction numbers in row
9 - 2 = 7
Count from 9 backward 2 times 8, 7
Subtract numbers

	a	b	c	d	e
# 1	0 – 0	1 – 0	2 – 0	3 - 0	4 – 0
# 2	5 – 0	6 – 0	7 – 0	8 – 0	9 - 0
# 3	1 – 1	2 – 1	3 – 1	4 – 1	5 - 1
# 4	6 – 1	7 – 1	8 – 1	9 – 1	8 – 7
# 5	2 – 2	3 – 2	4 – 2	5 – 2	6 – 2
# 6	7 – 2	8 – 2	9 – 2	9 – 8	9 - 7
# 7	3 – 3	4 – 3	5 – 3	6 – 3	7 – 3
# 8	4 – 4	5 – 4	6 – 4	7 – 4	8 – 4
# 9	5 – 5	6 – 5	7 – 5	8 – 5	9 – 5
# 10	6 – 6	7 – 6	8 – 6	9 – 6	9 - 9

Grade 1 Chapter 1 Lesson 4
Addition numbers in column 1

$$\begin{array}{r} 7 \\ +2 \\ \hline 9 \end{array} \qquad \begin{array}{r} 5 \\ +4 \\ \hline 9 \end{array}$$

Add numbers

	a	b	c	d	e
# 1	0 + 0	1 + 0	2 + 0	3 + 0	4 + 0
# 2	5 + 0	6 + 0	7 + 0	8 + 0	9 + 0
# 3	0 + 1	1 + 1	2 + 1	3 + 1	4 + 1
# 4	5 + 1	6 + 1	7 + 1	8 + 1	1 + 8
# 5	0 + 2	1 + 2	2 + 2	3 + 2	4 + 2
# 6	5 + 2	6 + 2	7 + 2	2 + 7	1 + 7
# 7	0 + 3	1 + 3	2 + 3	3 + 3	4 + 3
# 8	5 + 3	6 + 3	0 + 5	1 + 5	2 + 5
# 9	0 + 4	1 + 4	2 + 4	3 + 4	4 + 4
# 10	5 + 4	3 + 5	4 + 5	0 + 6	2 + 6

Grade 1 Chapter 1 Lesson 5
Addition numbers in column 2

$$\begin{array}{r} 8 \\ +1 \\ \hline 9 \end{array} \qquad \begin{array}{r} 6 \\ +3 \\ \hline 9 \end{array}$$

Add numbers

	a	b	c	d	e
# 1	$\begin{array}{r} 1 \\ +0 \\ \hline \end{array}$	$\begin{array}{r} 1 \\ +1 \\ \hline \end{array}$	$\begin{array}{r} 1 \\ +2 \\ \hline \end{array}$	$\begin{array}{r} 1 \\ +3 \\ \hline \end{array}$	$\begin{array}{r} 1 \\ +4 \\ \hline \end{array}$
# 2	$\begin{array}{r} 1 \\ +5 \\ \hline \end{array}$	$\begin{array}{r} 1 \\ +6 \\ \hline \end{array}$	$\begin{array}{r} 1 \\ +7 \\ \hline \end{array}$	$\begin{array}{r} 1 \\ +8 \\ \hline \end{array}$	$\begin{array}{r} 8 \\ +1 \\ \hline \end{array}$
# 3	$\begin{array}{r} 2 \\ +0 \\ \hline \end{array}$	$\begin{array}{r} 2 \\ +1 \\ \hline \end{array}$	$\begin{array}{r} 2 \\ +2 \\ \hline \end{array}$	$\begin{array}{r} 2 \\ +3 \\ \hline \end{array}$	$\begin{array}{r} 2 \\ +4 \\ \hline \end{array}$
# 4	$\begin{array}{r} 2 \\ +5 \\ \hline \end{array}$	$\begin{array}{r} 2 \\ +6 \\ \hline \end{array}$	$\begin{array}{r} 2 \\ +7 \\ \hline \end{array}$	$\begin{array}{r} 7 \\ +2 \\ \hline \end{array}$	$\begin{array}{r} 7 \\ +1 \\ \hline \end{array}$
# 5	$\begin{array}{r} 3 \\ +0 \\ \hline \end{array}$	$\begin{array}{r} 3 \\ +1 \\ \hline \end{array}$	$\begin{array}{r} 3 \\ +2 \\ \hline \end{array}$	$\begin{array}{r} 3 \\ +3 \\ \hline \end{array}$	$\begin{array}{r} 3 \\ +4 \\ \hline \end{array}$
# 6	$\begin{array}{r} 3 \\ +5 \\ \hline \end{array}$	$\begin{array}{r} 3 \\ +6 \\ \hline \end{array}$	$\begin{array}{r} 1 \\ +6 \\ \hline \end{array}$	$\begin{array}{r} 2 \\ +6 \\ \hline \end{array}$	$\begin{array}{r} 3 \\ +6 \\ \hline \end{array}$
# 7	$\begin{array}{r} 4 \\ +0 \\ \hline \end{array}$	$\begin{array}{r} 4 \\ +1 \\ \hline \end{array}$	$\begin{array}{r} 4 \\ +2 \\ \hline \end{array}$	$\begin{array}{r} 4 \\ +3 \\ \hline \end{array}$	$\begin{array}{r} 4 \\ +4 \\ \hline \end{array}$
# 8	$\begin{array}{r} 5 \\ +0 \\ \hline \end{array}$	$\begin{array}{r} 5 \\ +1 \\ \hline \end{array}$	$\begin{array}{r} 5 \\ +2 \\ \hline \end{array}$	$\begin{array}{r} 5 \\ +3 \\ \hline \end{array}$	$\begin{array}{r} 5 \\ +4 \\ \hline \end{array}$
# 9	$\begin{array}{r} 6 \\ +0 \\ \hline \end{array}$	$\begin{array}{r} 6 \\ +1 \\ \hline \end{array}$	$\begin{array}{r} 6 \\ +2 \\ \hline \end{array}$	$\begin{array}{r} 6 \\ +3 \\ \hline \end{array}$	$\begin{array}{r} 8 \\ +0 \\ \hline \end{array}$
# 10	$\begin{array}{r} 7 \\ +0 \\ \hline \end{array}$	$\begin{array}{r} 7 \\ +1 \\ \hline \end{array}$	$\begin{array}{r} 7 \\ +2 \\ \hline \end{array}$	$\begin{array}{r} 8 \\ +1 \\ \hline \end{array}$	$\begin{array}{r} 4 \\ +5 \\ \hline \end{array}$

Grade 1 Chapter 1 Lesson 6
Subtraction numbers in column

$$\begin{array}{r} 6 \\ \underline{-2} \\ 4 \end{array} \qquad \begin{array}{r} 7 \\ \underline{-2} \\ 5 \end{array}$$

Subtract numbers

	a	b	c	d	e
# 1	$\begin{array}{r} 0 \\ \underline{-0} \end{array}$	$\begin{array}{r} 1 \\ \underline{-0} \end{array}$	$\begin{array}{r} 2 \\ \underline{-0} \end{array}$	$\begin{array}{r} 3 \\ \underline{-0} \end{array}$	$\begin{array}{r} 4 \\ \underline{-0} \end{array}$
# 2	$\begin{array}{r} 5 \\ \underline{-0} \end{array}$	$\begin{array}{r} 6 \\ \underline{-0} \end{array}$	$\begin{array}{r} 7 \\ \underline{-0} \end{array}$	$\begin{array}{r} 8 \\ \underline{-0} \end{array}$	$\begin{array}{r} 9 \\ \underline{-0} \end{array}$
# 3	$\begin{array}{r} 1 \\ \underline{-1} \end{array}$	$\begin{array}{r} 2 \\ \underline{-1} \end{array}$	$\begin{array}{r} 3 \\ \underline{-1} \end{array}$	$\begin{array}{r} 4 \\ \underline{-1} \end{array}$	$\begin{array}{r} 5 \\ \underline{-1} \end{array}$
# 4	$\begin{array}{r} 6 \\ \underline{-1} \end{array}$	$\begin{array}{r} 7 \\ \underline{-1} \end{array}$	$\begin{array}{r} 8 \\ \underline{-1} \end{array}$	$\begin{array}{r} 9 \\ \underline{-1} \end{array}$	$\begin{array}{r} 8 \\ \underline{-7} \end{array}$
# 5	$\begin{array}{r} 2 \\ \underline{-2} \end{array}$	$\begin{array}{r} 3 \\ \underline{-2} \end{array}$	$\begin{array}{r} 4 \\ \underline{-2} \end{array}$	$\begin{array}{r} 5 \\ \underline{-2} \end{array}$	$\begin{array}{r} 6 \\ \underline{-2} \end{array}$
# 6	$\begin{array}{r} 7 \\ \underline{-2} \end{array}$	$\begin{array}{r} 8 \\ \underline{-2} \end{array}$	$\begin{array}{r} 9 \\ \underline{-2} \end{array}$	$\begin{array}{r} 9 \\ \underline{-8} \end{array}$	$\begin{array}{r} 9 \\ \underline{-7} \end{array}$
# 7	$\begin{array}{r} 3 \\ \underline{-3} \end{array}$	$\begin{array}{r} 4 \\ \underline{-3} \end{array}$	$\begin{array}{r} 5 \\ \underline{-3} \end{array}$	$\begin{array}{r} 6 \\ \underline{-3} \end{array}$	$\begin{array}{r} 7 \\ \underline{-3} \end{array}$
# 8	$\begin{array}{r} 4 \\ \underline{-4} \end{array}$	$\begin{array}{r} 5 \\ \underline{-4} \end{array}$	$\begin{array}{r} 6 \\ \underline{-4} \end{array}$	$\begin{array}{r} 7 \\ \underline{-4} \end{array}$	$\begin{array}{r} 8 \\ \underline{-4} \end{array}$
# 9	$\begin{array}{r} 5 \\ \underline{-5} \end{array}$	$\begin{array}{r} 6 \\ \underline{-5} \end{array}$	$\begin{array}{r} 7 \\ \underline{-5} \end{array}$	$\begin{array}{r} 8 \\ \underline{-5} \end{array}$	$\begin{array}{r} 9 \\ \underline{-5} \end{array}$
# 10	$\begin{array}{r} 6 \\ \underline{-6} \end{array}$	$\begin{array}{r} 7 \\ \underline{-6} \end{array}$	$\begin{array}{r} 8 \\ \underline{-6} \end{array}$	$\begin{array}{r} 9 \\ \underline{-6} \end{array}$	$\begin{array}{r} 9 \\ \underline{-9} \end{array}$

Grade 1 Chapter 1 Lesson 7
Test 1 A
Add or Subtract numbers in row

# 1	2 + 0	# 11	9 - 9
# 2	3 + 3	# 12	8 – 7
# 3	8 + 0	# 13	7 – 6
# 4	0 + 6	# 14	7 – 5
# 5	7 + 1	# 15	8 – 4
# 6	1 + 8	# 16	4 – 4
# 7	2 + 2	# 17	6 – 3
# 8	7 + 2	# 18	9 – 2
# 9	2 + 4	# 19	5 – 2
# 10	6 + 2	# 20	9 – 1

Add or Subtract numbers in column

# 21	1 + 4	# 22	6 + 2	# 23	3 + 3	# 24	3 + 4	# 25	5 + 3
# 26	3 + 6	# 27	2 + 4	# 28	4 + 3	# 29	4 + 4	# 30	4 + 5
# 31	9 -8	# 32	7 - 7	# 33	6 - 6	# 34	6 - 5	# 35	7 - 4
# 36	9 -3	# 37	5 - 3	# 38	8 - 2	# 39	4 - 2	# 40	8 - 1

Grade 1 Chapter 1 Lesson 7
Test 1 B
Add or Subtract numbers in row

1 1 + 6

2 4 + 5

3 1 + 5

4 5 + 2

5 3 + 5

6 5 + 4

7 0 + 6

8 6 + 1

9 2 + 6

10 6 + 3

11 8 - 8

12 9 – 6

13 9 – 5

14 5 – 5

15 6 – 4

16 8 – 3

17 4 – 3

18 7 – 2

19 3 – 2

20 7 – 0

Add or Subtract numbers in column

# 21 6 + 2	# 22 5 + 3	# 23 1 + 7	# 24 7 + 2	# 25 1 + 8
# 26 7 + 1	# 27 9 + 0	# 28 2 + 7	# 29 8 + 1	# 30 0 + 9
# 31 9 -7	# 32 8 - 6	# 33 8 - 5	# 34 9 - 4	# 35 5 - 4
# 36 7 -3	# 37 3 - 3	# 38 6 - 2	# 39 2 - 2	# 40 6 - 0

Grade 1 Chapter 1 Lesson 8
Money Skills

1 penny = 1 cent

1 nickel = 5 cents = 1c + 1c + 1c + 1c + 1c

1 dime = 10 cents = 5c + 5c

How many ones or pennies?
2 pennies = 1c + 1c
3 pennies = 1c + 1c + 1c
4 pennies = 1c + 1c + 1c + 1c
6 pennies = 5c + 1c
7 pennies = 5c + 1c + 1c
8 pennies = 5c + 1c + 1c + 1c
9 pennies = 5c + 1c + 1c + 1c + 1c

How many tens or dimes?
1 ten = 10 one
2 tens = 20 ones = 10c + 10c
3 tens = 30 ones = 10c + 10c + 10c
4 tens = 40 ones = 10c + 10c + 10c + 10c
5 tens = 50 ones = 10c + 10c + 10c + 10c + 10c
6 tens = 60 ones = 10c + 10c + 10c + 10c + 10c + 10c

Skip Counting. Write the missing numbers.
Count by 2 (+2)   1, 3, 5, ?, ?.

Grade 1 Answers Chapter 1

Lesson 1

	a	b	c	d	e
# 1	0 < 5	1 < 5	2 < 5	3 < 5	4 < 5
# 2	5 = 5	6 > 5	7 > 5	8 > 5	9 > 5
# 3	0 < 6	1 < 6	2 < 6	3 < 6	4 < 6
# 4	5 < 6	6 = 6	7 > 6	8 > 6	9 > 6
# 5	0 < 7	1 < 7	2 < 7	3 < 7	4 < 7
# 6	5 < 7	6 < 7	7 = 7	8 > 7	9 > 7
# 7	0 < 8	1 < 8	2 < 8	3 < 8	4 < 8
# 8	5 < 8	6 < 8	7 < 8	8 = 8	9 > 8
# 9	0 < 9	1 < 9	2 < 9	3 < 9	4 < 9
# 10	5 < 9	6 < 9	7 < 9	8 < 9	9 = 9

Lesson 2 Review 1 A

	a	b	c	d	e
# 1	0	1	2	3	4
# 2	5	6	7	8	9
# 3	1	2	3	4	5
# 4	6	7	8	9	9
# 5	2	3	4	5	6
# 6	7	8	9	9	8
# 7	3	4	5	6	7
# 8	8	9	5	6	7
# 9	4	5	6	7	8
# 10	9	8	9	6	8

Lesson 2 Review 1 B

	a	b	c	d	e
# 1	1	2	3	4	5
# 2	6	7	8	9	9
# 3	2	3	4	5	6
# 4	7	8	9	9	8
# 5	3	4	5	6	7
# 6	8	9	7	8	9
# 7	4	5	6	7	8
# 8	5	6	7	8	9
# 9	6	7	8	9	8
# 10	7	8	9	9	9

Grade 1 Answers Chapter 1

Lesson 3

	a	b	c	d	e
# 1	0	1	2	3	4
# 2	5	6	7	8	9
# 3	0	1	2	3	4
# 4	5	6	7	8	1
# 5	0	1	2	3	4
# 6	5	6	7	1	2
# 7	0	1	2	3	4
# 8	0	1	2	3	4
# 9	0	1	2	3	4
# 10	0	1	2	3	0

Lesson 4

	a	b	c	d	e
# 1	0	1	2	3	4
# 2	5	6	7	8	9
# 3	1	2	3	4	5
# 4	6	7	8	9	9
# 5	2	3	4	5	6
# 6	7	8	9	9	8
# 7	3	4	5	6	7
# 8	8	9	5	6	7
# 9	4	5	6	7	8
# 10	9	8	9	6	8

Lesson 5

	a	b	c	d	e
# 1	1	2	3	4	5
# 2	6	7	8	9	9
# 3	2	3	4	5	6
# 4	7	8	9	9	8
# 5	3	4	5	6	7
# 6	8	9	7	8	9
# 7	4	5	6	7	8
# 8	5	6	7	8	9
# 9	6	7	8	9	8
# 10	7	8	9	9	9

Lesson 6

	a	b	c	d	e
# 1	0	1	2	3	4
# 2	5	6	7	8	9
# 3	0	1	2	3	4
# 4	5	6	7	8	1
# 5	0	1	2	3	4
# 6	5	6	7	1	2
# 7	0	1	2	3	4
# 8	0	1	2	3	4
# 9	0	1	2	3	4
# 10	0	1	2	3	0

Lesson 7

Test 1 A

# 1	2	# 2	6	# 3	8	# 4	6	# 5	8
# 6	9	# 7	4	# 8	9	# 9	6	# 10	8
# 11	0	# 12	1	# 13	1	# 14	2	# 15	4
# 16	0	# 17	3	# 18	7	# 19	3	# 20	8
# 21	5	# 22	8	# 23	6	# 24	7	# 25	8
# 26	9	# 27	6	# 28	7	# 29	8	# 30	9
# 31	1	# 32	0	# 33	0	# 34	1	# 35	3
# 36	6	# 37	2	# 38	6	# 39	2	# 40	7

Lesson 7

Test 1 B

# 1	7	# 2	9	# 3	6	# 4	7	# 5	8
# 6	9	# 7	6	# 8	7	# 9	8	# 10	9
# 11	0	# 12	3	# 13	4	# 14	0	# 15	2
# 16	5	# 17	1	# 18	5	# 19	1	# 20	7
# 21	8	# 22	8	# 23	8	# 24	9	# 25	9
# 26	8	# 27	9	# 28	9	# 29	9	# 30	9
# 31	2	# 32	2	# 33	3	# 34	5	# 35	1
# 36	4	# 37	0	# 38	4	# 39	0	# 40	6

Lesson 8

Count by 2 (+2) 1, 3, 5, 7, 9.

Grade 1

Chapter 2

Two – Digit Numbers

Grade 1 Chapter 2 Contents
Two-Digit Numbers
Lesson 1 - 16
Basic Problems

Work with lesson 1, read rules and examples, 2) Solve the problems from contents, 3) Check answers in the end of chapter 1, 4) Work with the next lessons 2, 3, 4, 5, …, 5) Solve Test 1 A, 6) Grading Scale: A =10 points, B=9points, C=8 points, D=7 points, F=0-6 points, 7) If you solve 10 problems correctly, you get an A, 8) If have an A, you can move on to next chapter, 9) If your grade is lower than an A, work with the mistakes, then solve Test 1B, 10) When you finish all of the chapters, solve Final Exam in last chapter.

Lesson 1 One-digit numbers. Solve the problems # 16 - # 25
Lesson 2 Two-digit numbers. Rewrite and memorize numbers from 10 to 99.
Lesson 3 Comparison Numbers. Solve the problems # 15, # 20 (a, b, c, d, e)
Lesson 4 Comparison two-digit numbers. Solve the problems # 15, # 20 (a - e)
Lesson 5 Addition Numbers in row 1. Solve the problems # 15, # 20 (a - e)

Lesson 6 Addition Numbers in row 2. Solve the problems # 15, # 20 (a - e)
Lesson 7 Subtraction numbers in row. Solve the problems # 15, # 20 (a - e)
Lesson 8 Numbers 10 and 11. Solve the problems # 5, # 10 (a, b, c, d)
Lesson 9 Numbers 12 and 13. Solve the problems # 5, # 10 (a, b, c, d)
Lesson 10 Numbers 14. Solve the problems # 5, # 10 (a, b, c, d)

Lesson 11 Numbers 15. Solve the problems # 5, # 10 (a, b, c, d)
Lesson 12 Numbers 16. Solve the problems # 5, # 10 (a, b, c, d)
Lesson 13 Numbers 17. Solve the problems # 5, # 10 (a, b, c, d)
Lesson 14 Numbers 18. Solve the problems # 5, #10 (a, b, c, d)
Lesson 15 Numbers 19. Solve the problems # 5, #10 (a, b, c, d)

Lesson 16 Test 2 for 50 minutes (10 problems)
Solve Test 2 A # 1, 5, 9, 13, 17, 21, 25, 29, 33, 37
Solve Test 2 B # 1, 5, 9, 13, 17, 21, 25, 29, 33, 37

Answers Chapter 2

Grade 1 Chapter 2 Lesson 1

One-digit numbers

One-digit numbers have one digit

This is numbers from 0 to 9

0, 1, 2, 3, 4, 5, 6, 7, 8, 9

Add or Subtract numbers in column

# 1	1	# 2	6	# 3	3	# 4	3	# 5	5
	+4		+2		+3		+4		+ 3
# 6	3	# 7	2	# 8	4	# 9	4	# 10	4
	+6		+ 4		+ 3		+ 4		+5
# 11	9	# 12	7	# 13	6	# 14	6	# 15	7
	-8		- 7		- 6		-5		-4
# 16	9	# 17	5	# 18	8	# 19	4	# 20	8
	-3		- 3		- 2		-2		- 1

Count money: 50 cents, 1 dollar

1 cent = 1 penny

1 nickel = 5 cent

1 dime = 10 cent

1 quarter = 25 cent

Add or Subtract numbers in column

# 21	$6	# 22	$5	# 23	$1	# 24	$7	# 25	$1
	+ $2		+$3		+ $7		+ $2		+ $8
# 26	$7	# 27	$9	# 28	$2	# 29	$8	# 30	$ 0
	+ $1		+ $0		+ $7		+ $1		+$ 9
# 31	$9	#32	$8	# 33	$8	# 34	$ 9	# 35	$5
	-$7		- $6		- $5		- $4		- $4
# 36	$7	# 37	$3	# 38	$6	# 39	$2	# 40	$6
	-$3		- $3		- $2		- $2		- $0

Grade 1 Chapter 2 Lesson 2
Two-digit numbers

Two - digit numbers have two digits
This is numbers from 10 to 99

Digit on the left are called tens
Digit on the right are called ones
Digits 0, 1, 2, 3, 4, 5, 6, 7, 8, 9

Ten, eleven, twelve, thirteen, fourteen,
fifteen, sixteen, seventeen, eighteen, nineteen

Read numbers 10 to 19: 10, 11, 12, 13, 14, 15, 16, 17, 18, 19
Read numbers 20 through 29: 20, 21, 22, 23, 24, 25, 26, 27, 28, 29
Read numbers 30 through 39: 30, 31, 32, 33, 34, 35, 36, 37, 38, 39
Read numbers 40 through 49: 40, 41, 42, 43, 44, 45, 46, 47, 48, 49
Read numbers 50 through 59: 50, 51, 52, 53, 54, 55, 56, 57, 58, 59
Read numbers 60 through 69: 60, 61, 62, 63, 64, 65, 66, 67, 68, 69
Read numbers 70 through 79: 70, 71, 72, 73, 74, 75, 76, 77, 78, 79
Read numbers 80 through 89: 80, 81, 82, 83, 84, 85, 86, 87, 88, 89
Read numbers 90 through 99: 90, 91, 92, 93, 94, 95, 96, 97, 98, 99

Rewrite and Memorize numbers from 10 to 99

Three-digit numbers

Three-digit numbers content three digits. This is numbers 100, 101, ... 999
Least three-digit number: 100. Great three-digit number: 999
Digits for three-digit numbers from right to the left: Ones, tens, and hundreds: 238
Counting numbers are numbers that use for calculations 1, 2, 3, 4, . . . (+ 1)
Odd numbers are 1, 3, 5, 7, 9, 11, 13, 15, 17, 19... (+ 2)
Even numbers are 0, 2, 4, 6, 8, 10, 12, 14, 16, 18, 20 ... (+ 2)

Remember

10 = 9 + 1 = 8 + 2 = 7 + 3 = 6 + 4 = 5 + 5	15 = 9 + 6 = 8 + 7
11 = 9 + 2 = 8 + 3 = 7 + 4 = 6 + 5	16 = 9 + 7 = 8 + 8
12 = 9 + 3 = 8 + 4 = 7 + 5 = 6 + 6	17 = 9 + 8
13 = 9 + 4 = 8 + 5 = 7 + 6	18 = 9 + 9
14 = 9 + 5 = 8 + 6 = 7 + 7	19 = 10 + 9

Grade 1 Chapter 2 Lesson 3
Comparison numbers

To compare numbers, find what number greater or smaller or numbers are equal. One-digit numbers always less than two-digit numbers 9 < 11

Two-digit numbers always greater than one-digit numbers 19 > 8

Put symbols < or > or = instead of question mark

	a	b	c	d	e
# 1	0 ? 10	10 ? 1	2 ? 10	10 ? 3	4 ? 10
# 2	0 ? 11	11 ? 1	2 ? 11	11 ? 3	11 ? 4
# 3	0 ? 12	12 ? 1	2 ? 12	3 ? 12	12 ? 4
# 4	13 ? 0	1 ? 13	13 ? 2	13 ? 3	4 ? 13
# 5	14 ? 0	14 ? 1	2 ? 14	3 ? 14	4 ? 14
# 6	15 ? 0	1 ? 15	15 ? 2	3 ? 15	15 ? 4
# 7	0 ? 16	1 ? 16	16 ? 2	16 ? 3	4 ? 16
# 8	17 ? 0	1 ? 17	17 ? 2	17 ? 3	4 ? 17
# 9	18 ? 0	1 ? 18	2 ? 18	18 ? 3	4 ? 18
# 10	19 ? 0	1 ? 19	19 ? 2	19 ? 3	4 ? 19
# 11	5 ? 10	10 ? 6	7 ? 10	10 ? 8	9 ? 10
# 12	11 ? 5	6 ? 11	11 ? 7	11? 8	9 ? 11
# 13	5 ? 12	6 ? 12	12 ? 7	8 ? 12	9 ? 12
# 14	13 ? 5	6 ? 13	13 ? 7	13 ? 8	13 ? 9
# 15	5 ? 14	14 ? 6	7 ? 14	8 ? 14	9 ? 14
# 16	15 ? 5	6 ? 15	15 ? 7	8 ? 15	15 ? 9
# 17	5 ? 16	16 ? 6	7 ? 16	8 ? 16	16 ? 9
# 18	17 ? 5	17 ? 6	7 ? 17	17 ? 8	17 ? 9
# 19	5 ? 18	18 ? 6	7 ? 18	18 ? 8	9 ? 18
# 20	19 ? 5	6 ? 19	19 ? 7	8 ? 19	19 ? 9

Grade 1 Chapter 2 Lesson 4
Comparison two-digit numbers

To compare numbers, find what number greater or smaller or numbers equal.

Compare the tens and then compare the ones 12 < 18   18 > 12

Put symbols < or > or = instead of question mark

	a	b	c	d	e
# 1	10 ? 10	11 ? 11	12 ? 12	11 ? 10	12 ? 10
# 2	10 ? 11	12 ? 11	12 ? 13	11 ? 12	13 ? 12
# 3	12 ? 10	11 ? 13	14 ? 12	13 ? 11	12 ? 14
# 4	10 ? 13	14 ? 11	15 ? 12	11 ? 14	12 ? 15
# 5	14 ? 10	11 ? 15	16 ? 12	15 ? 11	12 ? 16
# 6	10 ? 15	16 ? 11	12 ? 17	11 ? 16	17 ? 12
# 7	10 ? 16	17 ? 11	12 ? 18	11 ? 17	18 ? 12
# 8	17 ? 10	11 ? 18	19 ? 12	18 ? 11	12 ? 19
# 9	10 ? 18	11 ? 19	19 ? 10	18 ? 10	10 ? 19
# 10	11 ? 17	18 ? 15	12 ? 16	15 ? 18	16 ? 12
# 11	13 ? 13	14 ? 14	15 ? 15	14 ? 13	13 ? 12
# 12	13 ? 14	15 ? 14	15 ? 16	14 ? 15	16 ? 15
# 13	15 ? 13	14 ? 16	17 ? 15	16 ? 14	15 ? 17
# 14	13 ? 16	17 ? 14	15 ? 18	14 ? 17	18 ? 15
# 15	17 ? 13	14 ? 18	15 ? 19	18 ? 14	19 ? 15
# 16	13 ? 18	19 ? 14	13 ? 19	14 ? 19	19 ? 13
# 17	16 ? 16	17 ? 17	18 ? 18	17 ? 16	18 ? 13
# 18	16 ? 17	18 ? 17	18 ? 19	17 ? 18	19 ? 18
# 19	18 ? 16	17 ? 19	19 ? 19	19 ? 17	16 ? 18
# 20	16 ? 19	19 ? 18	15 ? 17	18 ? 19	17 ? 15

Grade 1 Chapter 2 Lesson 5

Addition numbers in row 1

5 + 9 = 9 + 5 = 14

Count from 9 forward 5 times 10, 11, 12, 13, 14

4 + 10 = 10 + 4 = 14

Count from 10 forward 4 times 11, 12, 13, 14

Add numbers

	a	b	c	d	e
# 1	0 + 1	1 + 1	2 + 1	3 + 1	4 + 1
# 2	5 + 1	6 + 1	7 + 1	8 + 1	9 + 1
# 3	0 + 2	1 + 2	2 + 2	3 + 2	4 + 2
# 4	5 + 2	6 + 2	7 + 2	8 + 2	9 + 2
# 5	0 + 3	1 + 3	2 + 3	3 + 3	4 + 3
# 6	5 + 3	6 + 3	7 + 3	8 + 3	9 + 3
# 7	0 + 4	1 + 4	2 + 4	3 + 4	4 + 4
# 8	5 + 4	6 + 4	7 + 4	8 + 4	9 + 4
# 9	0 + 5	1 + 5	2 + 5	3 + 5	4 + 5
# 10	5 + 5	6 + 5	7 + 5	8 + 5	9 + 5
# 11	0 + 6	1 + 6	2 + 6	3 + 6	4 + 6
# 12	5 + 6	6 + 6	7 + 6	8 + 6	9 + 6
# 13	0 + 7	1 + 7	2 + 7	3 + 7	4 + 7
# 14	5 + 7	6 + 7	7 + 7	8 + 7	9 + 7
# 15	0 + 8	1 + 8	2 + 8	3 + 8	4 + 8
# 16	5 + 8	6 + 8	7 + 8	8 + 8	9 + 8
# 17	0 + 9	1 + 9	2 + 9	3 + 9	4 + 9
# 18	5 + 9	6 + 9	7 + 9	8 + 9	9 + 9
# 19	0 + 10	1 + 10	2 + 10	3 + 10	4 + 10
# 20	5 + 10	6 + 10	7 + 10	8 + 10	9 + 10

Grade 1 Chapter 2 Lesson 6

Addition numbers in row 2

7 + 12 = 12 + 7 = 10 + 2 + 7 = 10 + 9 = 19
6 + 12 = 12 + 6 = 10 + 2 + 6 = 10 + 8 = 18

Add numbers

	a	b	c	d	e
# 1	8 + 0	8 + 1	8 + 2	8 + 3	8 + 4
# 2	8 + 5	8 + 6	8 + 7	8 + 8	8 + 9
# 3	9 + 0	9 + 1	9 + 2	9 + 3	9 + 4
# 4	9+ 5	9 + 6	9 + 7	9 + 8	9 + 9
# 5	10 + 0	10 + 1	10 + 2	10 + 3	10 + 4
# 6	10 + 5	10 + 6	10 + 7	10 + 8	10 + 9
# 7	11 + 0	11 + 1	11 + 2	11 + 3	11 + 4
# 8	11 + 5	11 + 6	11 + 7	11 + 8	8 + 11
# 9	12 + 0	12 + 1	12 + 2	12 + 3	12 + 4
# 10	12 + 5	12 + 6	12 + 7	7 + 12	6 + 12
# 11	13 + 0	13 + 1	13 + 2	13 + 3	13 + 4
# 12	13 + 5	13 + 6	4 + 13	5 + 13	6 + 13
# 13	14 + 0	14 + 1	14 + 2	14 + 3	14 + 4
# 14	14 + 5	2 + 14	3 + 14	4 + 14	5 + 14
# 15	15 + 0	15 + 1	15 + 2	15 + 3	15 + 4
# 16	0 + 15	1 + 15	2 + 15	3 + 15	4 + 15
# 17	16 + 0	16 + 1	16 + 2	16 + 3	19 + 0
# 18	0 + 16	1 + 16	2 + 16	3 + 16	0 + 19
# 19	17 + 0	17 + 1	17 + 2	18 + 0	18 + 1
# 20	0 + 17	1 + 17	2 + 17	0 + 18	1 + 18

Grade 1 Chapter 2 Lesson 7

Subtraction numbers in row

14 – 4 = 10

Count from 14 backward 4 times 13, 12, 11, 10

18 – 4 = 14

Count from 18 backward 4 times 17, 16, 15, 14

Subtract numbers

	a	b	c	d	e
# 1	10 - 0	10 - 1	10 - 2	10 - 3	10 - 4
# 2	10 - 5	10 - 6	10 - 7	10 - 8	10 - 9
# 3	11 - 0	11 - 1	11 - 2	11 - 3	11 - 4
# 4	11 - 5	11 - 6	11 - 7	11 - 8	11 - 9
# 5	12 - 0	12 - 1	12 - 2	12 - 3	12 - 4
# 6	12 - 5	12 - 6	12 - 7	12 - 8	12 - 9
# 7	13 - 0	13 - 1	13 - 2	13 - 3	13 - 4
# 8	13 - 5	13 - 6	13 - 7	13 - 8	13 - 9
# 9	14 - 0	14 - 1	14 - 2	14 - 3	14 - 4
# 10	14 - 5	14 - 6	14 - 7	14 - 8	14 - 9
# 11	15 – 0	15 – 1	15 - 2	15 - 3	15 - 4
# 12	15 - 5	15 - 6	15 - 7	15 - 8	15 - 9
# 13	16 - 0	16 - 1	16 - 2	16 - 3	16 - 4
# 14	16 - 5	16 - 6	16 - 7	16 - 8	16 - 9
# 15	17 - 0	17 - 1	17 - 2	17 - 3	17 - 4
# 16	17 - 5	17 - 6	17 - 7	17 - 8	17 - 9
# 17	18 - 0	18 - 1	18 - 2	18 - 3	18 - 4
# 18	18 - 5	18 - 6	18 - 7	18 - 8	18 - 9
# 19	19 - 0	19 - 1	19 - 2	19 - 3	19 - 4
# 20	19 - 5	19 - 6	19 - 7	19 - 8	19 - 9

Grade 1 Chapter 2 Lesson 8

Addition and Subtraction in column
Number 10 and Number 11

6	11
+4	-2
10	9

Add or Subtract numbers

	a	b	c	d
# 1	10 + 0	10 - 1	5 + 5	10 - 6
# 2	9 + 1	10 - 2	4 + 6	10 - 7
# 3	8 + 2	10 - 3	3 + 7	10 - 8
# 4	7 + 3	10 - 4	2 + 8	10 - 9
# 5	6 + 4	10 - 5	1 + 9	10 - 10
# 6	10 + 1	11 - 1	5 + 6	11 - 6
# 7	9 + 2	11 - 2	4 + 7	11 - 7
# 8	8 + 3	11 - 3	3 + 8	11 - 8
# 9	7 + 4	11 - 4	2 + 9	11 - 9
# 10	6 + 5	11 - 5	1 + 10	11 - 10

Grade 1 Chapter 2 Lesson 9
Addition and Subtraction in column
Number 12 and Number 13

9	13
+3	-7
12	6

Add or Subtract numbers

	a	b	c	d
# 1	12 + 0	12 - 0	7 + 5	12 - 5
# 2	11 + 1	12 - 1	6 + 6	12 - 6
# 3	10 + 2	12 - 2	5 + 7	12 - 7
# 4	9 + 3	12 - 3	4 + 8	12 - 8
# 5	8 + 4	12 - 4	3 + 9	12 - 9
# 6	13 + 0	13 - 0	8 + 5	13 - 5
# 7	12 + 1	13 - 1	7 + 6	13 - 6
# 8	11 + 2	13 - 2	6 + 7	13 - 7
# 9	10 + 3	13 - 3	5 + 8	13 - 8
# 10	9 + 4	13 - 4	4 + 9	13 - 9

Grade 1 Chapter 2 Lesson 10
Addition and Subtraction in column
Number 14

9	14
+ 5	-9
14	5

Add or Subtract numbers

	a	b	c	d
# 1	14 + 0	14 - 0	9 + 5	14 - 5
# 2	13 + 1	14 - 1	8 + 6	14 - 6
# 3	12 + 2	14 - 2	7 + 7	14 - 7
# 4	11 + 3	14 - 3	6 + 8	14 - 8
# 5	10 + 4	14 - 4	5 + 9	14 - 9
# 6	4 + 10	14 - 10	12 + 0	12 - 0
# 7	3 + 11	14 - 11	12 + 0	12 - 0
# 8	2 + 12	14 - 12	11 + 1	12 - 9
# 9	1 + 13	14 - 13	10 + 2	12 - 10
# 10	0 + 14	14 - 14	12 - 12	12 - 11

Grade 1 Chapter 2 Lesson 11
Addition and Subtraction in column
Number 15

9	15
+6	-9
15	6

Add or Subtract numbers

	a	b	c	d
# 1	15 + 0	15 - 0	10 + 5	15 - 5
# 2	14 + 1	15 - 1	9 + 6	15 - 6
# 3	13 + 2	15 - 2	8 + 7	15 - 7
# 4	12 + 3	15 - 3	7 + 8	15 - 8
# 5	11 + 4	15 - 4	6 + 9	15 - 9
# 6	5 + 10	15 - 10	13 + 0	13 - 10
# 7	4 + 11	15 - 11	12 + 1	13 - 11
# 8	3 + 12	15 - 12	11 + 2	13 - 12
# 9	2 + 13	15 - 13	10 + 3	13 - 13
# 10	1 + 14	15 - 14	0 + 13	15 - 15

Grade 1 Chapter 2 Lesson 12
Addition and Subtraction in column
Number 16

9	16
+7	-9
16	7

Add or Subtract numbers

	a	b	c	d
# 1	16 + 0	16 - 0	11 + 5	16 - 5
# 2	15 + 1	16 - 1	10 + 6	16 - 6
# 3	14 + 2	16 - 2	9 + 7	16 - 7
# 4	13 + 3	16 - 3	8 + 8	16 - 8
# 5	12 + 4	16 - 4	7 + 9	16 - 9
# 6	6 + 10	16 - 10	1 + 15	16 - 5
# 7	5 + 11	16 - 11	0 + 16	16 - 6
# 8	4 + 12	16 - 12	9 + 7	16 - 7
# 9	3 + 13	16 - 13	8 + 8	16 - 8
# 10	2 + 14	16 - 14	7 + 9	16 - 9

Grade 1 Chapter 2 Lesson 13
Addition and Subtraction in column
Number 17

9	17
+8	-9
17	8

Add or Subtract numbers

	a	b	c	d
# 1	17 + 0	17 - 0	12 + 5	17 - 5
# 2	16 + 1	17 - 1	11 + 6	17 - 6
# 3	15 + 2	17 - 2	10 + 7	17 - 7
# 4	14 + 3	17 - 3	9 + 8	17 - 8
# 5	13 + 4	17 - 4	8 + 9	17 - 9
# 6	7 + 10	17 - 10	2 + 15	17 - 5
# 7	6 + 11	17 - 11	1 + 16	17 - 6
# 8	5 + 12	17 - 12	0 + 17	17 - 7
# 9	4 + 13	17 - 13	9 + 8	17 - 8
# 10	3 + 14	17 - 14	8 + 9	17 - 9

Grade 1 Chapter 2 Lesson 14
Addition and Subtraction in column
Number 18

9	18
+9	-9
18	9

Add or Subtract numbers

	a	b	c	d
# 1	18 + 0	18 - 0	13 + 5	18 - 5
# 2	17 + 1	18 - 1	12 + 6	18 - 6
# 3	16 + 2	18 - 2	11 + 7	18 - 7
# 4	15 + 3	18 - 3	10 + 8	18 - 8
# 5	14 + 4	18 - 4	9 + 9	18 - 9
# 6	8 + 10	18 - 10	3 + 15	18 - 5
# 7	7 + 11	18 - 11	2 + 16	18 - 6
# 8	6 + 12	18 - 12	1 + 17	18 - 7
# 9	5 + 13	18 - 13	0 + 18	18 - 8
# 10	4 + 14	18 - 14	9 + 9	18 - 9

Grade 1 Chapter 2 Lesson 15
Addition and Subtraction in column
Number 19

10	19
+9	-10
19	9

Add or Subtract numbers

	a	b	c	d
# 1	19 + 0	19 - 0	14 + 5	19 - 5
# 2	18 + 1	19 - 1	13 + 6	19 - 6
# 3	17 + 2	19 - 2	12 + 7	19 - 7
# 4	16 + 3	19 - 3	11 + 8	19 - 8
# 5	15 + 4	19 - 4	10 + 9	19 - 9
# 6	9 + 10	19 - 10	4 + 15	19 - 15
# 7	8 + 11	19 - 11	3 + 16	19 - 16
# 8	7 + 12	19 - 12	2 + 17	19 - 17
# 9	6 + 13	19 - 13	1 + 18	19 - 18
# 10	5 + 14	19 - 14	0 + 19	19 - 19

Grade 1 Chapter 2 Lesson 16
Test 2 A
Add or Subtract numbers in row

# 1	8 + 5	# 11	13 - 4
# 2	6 + 8	# 12	13 – 6
# 3	8 + 7	# 13	13 – 8
# 4	8 + 8	# 14	14 – 5
# 5	9 + 8	# 15	14 – 7
# 6	9 + 5	# 16	15 – 6
# 7	6 + 9	# 17	15 – 8
# 8	9 + 7	# 18	16 – 7
# 9	8 + 9	# 19	17 – 8
# 10	9 + 9	# 20	16 – 9

Add or Subtract numbers in column

# 21	9 + 9	# 22	8 + 9	# 23	9 + 8	# 24	7 + 9	# 25	8 + 8
# 26	9 + 7	# 27	6 + 9	# 28	8 + 7	# 29	7 + 8	# 30	9 + 6
# 31	18 -9	# 32	17 - 9	# 33	16 - 8	# 34	15 - 6	# 35	15 - 8
# 36	14 -5	# 37	14 - 7	# 38	14 - 9	# 39	13 - 6	# 40	13 - 8

Grade 1 Chapter 2 Lesson 16
Test 2 B
Add or Subtract numbers in row

1 5 + 8
2 8 + 6
3 7 + 8
4 8 + 8
5 8 + 9

6 5 + 9
7 9 + 6
8 7 + 9
9 9 + 8
10 9 + 9

11 13 - 5
12 13 – 7
13 13 – 9
14 14 – 6
15 14 – 8

16 15 – 7
17 15 – 9
18 16 – 8
19 17 – 9
20 18 – 9

Add or Subtract numbers in column

# 21	9	# 22	8	# 23	7	# 24	6	# 25	5
	+ 5		+ 6		+ 7		+ 8		+ 9
# 26	8	# 27	7	# 28	6	# 29	5	# 30	4
	+ 5		+ 6		+ 7		+ 8		+ 9
# 31	17	# 32	16	# 33	16	# 34	15	# 35	15
	-8		- 7		- 9		- 7		- 9
# 36	14	# 37	14	# 38	13	# 39	13	# 40	13
	-6		- 8		- 5		- 7		- 9

Grade 1 Answers Chapter 2

Lesson 1

# 1	5	# 2	8	# 3	6	# 4	7	# 5	8
# 6	9	# 7	6	# 8	7	# 9	8	# 10	9
# 11	1	# 12	0	# 13	0	# 14	1	# 15	3
# 16	6	# 17	2	# 18	6	# 19	2	# 20	7
# 21	8	# 22	8	# 23	8	# 24	9	# 25	9
# 26	8	# 27	9	# 28	9	# 29	9	# 30	9
# 31	2	# 32	2	# 33	3	# 34	5	# 35	1
# 36	4	# 37	0	# 38	4	# 39	0	# 40	6

Lesson 3

	a	b	c	d	e
# 1	<	>	<	>	<
# 2	<	>	<	>	>
# 3	<	>	<	<	>
# 4	>	<	>	>	<
# 5	>	>	<	<	<
# 6	>	<	>	<	>
# 7	<	<	>	>	<
# 8	>	<	>	>	<
# 9	>	<	<	>	<
# 10	>	<	>	>	<
# 11	<	>	<	>	<
# 12	>	<	>	>	<
# 13	<	<	>	<	<
# 14	>	<	>	>	>
# 15	<	>	<	<	<
# 16	>	<	>	<	>
# 17	<	>	<	<	>
# 18	>	>	<	>	>
# 19	<	>	<	>	<
# 20	>	<	>	<	>

Grade 1 Answers Chapter 2

Lesson 4

	a	b	c	d	e
# 1	=	=	=	>	>
# 2	<	>	<	<	>
# 3	>	<	>	>	<
# 4	<	>	>	<	<
# 5	>	<	>	>	<
# 6	<	>	<	<	>
# 7	<	>	<	<	>
# 8	>	<	>	>	<
# 9	<	<	>	>	<
# 10	<	>	<	<	>
# 11	=	=	=	>	>
# 12	<	>	<	<	>
# 13	>	<	>	>	<
# 14	<	>	<	<	>
# 15	>	<	<	>	>
# 16	<	>	<	<	>
# 17	=	=	=	>	>
# 18	<	>	<	<	>
# 19	>	<	=	>	<
# 20	<	>	<	<	>

Lesson 5

	a	b	c	d	e
# 1	1	2	3	4	5
# 2	6	7	8	9	10
# 3	2	3	4	5	6
# 4	7	8	9	10	11
# 5	3	4	5	6	7
# 6	8	9	10	11	12
# 7	4	5	6	7	8
# 8	9	10	11	12	13
# 9	5	6	7	8	9
# 10	10	11	12	13	14

Grade 1 Answers Chapter 2

Lesson 5

	a	b	c	d	e
# 11	6	7	8	9	10
# 12	11	12	13	14	15
# 13	7	8	9	10	11
# 14	12	13	14	15	16
# 15	8	9	10	11	12
# 16	13	14	15	16	17
# 17	9	10	11	12	13
# 18	14	15	16	17	18
# 19	10	11	12	13	14
# 20	15	16	17	18	19

Lesson 6

	a	b	c	d	e
# 1	8	9	10	11	12
# 2	13	14	15	16	17
# 3	9	10	11	12	13
# 4	14	15	16	17	18
# 5	10	11	12	13	14
# 6	15	16	17	18	19
# 7	11	12	13	14	15
# 8	16	17	18	19	19
# 9	12	13	14	15	16
# 10	17	18	19	19	18
# 11	13	14	15	16	17
# 12	18	19	17	18	19
# 13	14	15	16	17	18
# 14	19	16	17	18	19
# 15	15	16	17	18	19
# 16	15	16	17	18	19
# 17	16	17	18	19	19
# 18	16	17	18	19	19
# 19	17	18	19	18	19
# 20	17	18	19	18	19

Grade 1 Answers Chapter 2

Lesson 7

	a	b	c	d	e
# 1	10	9	8	7	6
# 2	5	4	3	2	1
# 3	11	10	9	8	7
# 4	6	5	4	3	2
# 5	12	11	10	9	8
# 6	7	6	5	4	3
# 7	13	12	11	10	9
# 8	8	7	6	5	4
# 9	14	13	12	11	10
# 10	9	8	7	6	5
# 11	15	14	13	12	11
# 12	10	9	8	7	6
# 13	16	15	14	13	12
# 14	11	10	9	8	7
# 15	17	16	15	14	13
# 16	12	11	10	9	8
# 17	18	17	16	15	14
# 18	13	12	11	10	9
# 19	19	18	17	16	15
# 20	14	13	12	11	10

Lesson 8

	a	b	c	d
# 1	10	9	10	4
# 2	10	8	10	3
# 3	10	7	10	2
# 4	10	6	10	1
# 5	10	5	10	0
# 6	11	10	11	5
# 7	11	9	11	4
# 8	11	8	11	3
# 9	11	7	11	2
# 10	11	6	11	1

Lesson 9

	a	b	c	d
# 1	12	12	12	7
# 2	12	11	12	6
# 3	12	10	12	5
# 4	12	9	12	4
# 5	12	8	12	3
# 6	13	13	13	8
# 7	13	12	13	7
# 8	13	11	13	6
# 9	13	10	13	5
# 10	13	9	13	4

Lesson 10

	a	b	c	d
# 1	14	14	14	9
# 2	14	13	14	8
# 3	14	12	14	7
# 4	14	11	14	6
# 5	14	10	14	5
# 6	14	4	12	12
# 7	14	3	12	12
# 8	14	2	12	3
# 9	14	1	12	2
# 10	14	0	0	1

Lesson 11

	a	b	c	d
# 1	15	15	15	10
# 2	15	14	15	9
# 3	15	13	15	8
# 4	15	12	15	7
# 5	15	11	15	6
# 6	15	5	13	3
# 7	15	4	13	2
# 8	15	3	13	1
# 9	15	2	13	0
# 10	15	1	13	0

Grade 1 Answers Chapter 2

Lesson 12

	a	b	c	d
# 1	16	16	16	11
# 2	16	15	16	10
# 3	16	14	16	9
# 4	16	13	16	8
# 5	16	12	16	7
# 6	16	6	16	11
# 7	16	5	16	10
# 8	16	4	16	9
# 9	16	3	16	8
# 10	16	2	16	7

Lesson 13

	a	b	c	d
# 1	17	17	17	12
# 2	17	16	17	11
# 3	17	15	17	10
# 4	17	14	17	9
# 5	17	13	17	8
# 6	17	7	17	12
# 7	17	6	17	11
# 8	17	5	17	10
# 9	17	4	17	9
# 10	17	3	17	8

Lesson 14

	a	b	c	d
# 1	18	18	18	13
# 2	18	17	18	12
# 3	18	16	18	11
# 4	18	15	18	10
# 5	18	14	18	9
# 6	18	8	18	13
# 7	18	7	18	12
# 8	18	6	18	11
# 9	18	5	18	10
# 10	18	4	18	9

Grade 1 Answers Chapter 2

Lesson 15

	a	b	c	d
# 1	19	19	19	14
# 2	19	18	19	13
# 3	19	17	19	12
# 4	19	16	19	11
# 5	19	15	19	10
# 6	19	9	19	4
# 7	19	8	19	3
# 8	19	7	19	2
# 9	19	6	19	1
# 10	19	5	19	0

Lesson 16

Test 2 A

# 1	13	# 2	14	# 3	15	# 4	16	# 5	17
# 6	14	# 7	15	# 8	16	# 9	17	# 10	18
# 11	9	# 12	7	# 13	5	# 14	9	# 15	7
# 16	9	# 17	7	# 18	9	# 19	9	# 20	7
# 21	18	# 22	17	# 23	17	# 24	16	# 25	16
# 26	16	# 27	15	# 28	15	# 29	15	# 30	15
# 31	9	# 32	8	# 33	8	# 34	9	# 35	7
# 36	9	# 37	7	# 38	5	# 39	7	# 40	5

Test 2 B

# 1	13	# 2	14	# 3	15	# 4	16	# 5	17
# 6	14	# 7	15	# 8	16	# 9	17	# 10	18
# 11	8	# 12	6	# 13	4	# 14	8	# 15	6
# 16	8	# 17	6	# 18	8	# 19	8	# 20	9
# 21	14	# 22	14	# 23	14	# 24	14	# 25	14
# 26	13	# 27	13	# 28	13	# 29	13	# 30	13
# 31	9	# 32	9	# 33	7	# 34	8	# 35	6
# 36	8	# 37	6	# 38	8	# 39	6	# 40	4

Grade 1

Chapter 3

Three

Numbers

Grade 1 Chapter 3 Contents
Three Numbers
Lesson 1 - 12
Basic Problems

Work with lesson 1, read rules and examples, 2) Solve the problems from contents, 3) Check answers in the end of chapter 1, 4) Work with the next lessons 2, 3, 4, 5, ..., 5) Solve Test 1 A, 6) Grading Scale: A =10 points, B=9points, C=8 points, D=7 points, F=0-6 points, 7) If you solve 10 problems correctly, you get an A, 8) If have an A, you can move on to next chapter, 9) If your grade is lower than an A, work with the mistakes, then solve Test 1B, 10) When you finish all of the chapters, solve Final Exam in last chapter.

Grade 1 Chapter 3 Lesson 1
Three Numbers
Basic Problems
Addition three numbers in row 1
<u>Three numbers can add in any order</u>

9 + 4 + 3 = 13 + 3 = 16

9 + 4 + 3 = 9 + 7 = 16

9+3+4=12+4=16

9+3+4=9+7=16

Add numbers

	a	b	c	d
# 1	0 + 1 + 0	0 + 2 + 0	0 + 3 + 0	0 + 4 + 0
# 2	1 + 1 + 1	1 + 2 + 1	1 + 3 + 1	1 + 4 + 1
# 3	2 + 1 + 2	2 + 2 + 2	2 + 3 + 2	2 + 4 + 2
# 4	3 + 1 + 3	3 + 2 + 3	3 + 3 + 3	3 + 4 + 3
# 5	4 + 1 + 4	4 + 2 + 4	4 + 3 + 4	4 + 4 + 4
# 6	5 + 1 + 5	5 + 2 + 5	5 + 3 + 5	5 + 4 + 5
# 7	6 + 1 + 6	6 + 2 + 6	6 + 3 + 6	6 + 4 + 6
# 8	7 + 1 + 7	7 + 2 + 7	7 + 3 + 7	7 + 4 + 7
# 9	8 + 1 + 8	8 + 2 + 8	8 + 3 + 8	2 + 9 + 2
# 10	9 + 1 + 9	0 + 9 + 0	1 + 9 + 1	3 + 9 + 3
# 11	0 + 5 + 0	0 + 6 + 0	0 + 7 + 0	0 + 8 + 0
# 12	1 + 5 + 1	1 + 6 + 1	1 + 7 + 1	1 + 8 + 1
# 13	2 + 5 + 2	2 + 6 + 2	2 + 7 + 2	2 + 8 + 2
# 14	3 + 5 + 3	3 + 6 + 3	3 + 7 + 3	3 + 8 + 3
# 15	4 + 5 + 4	4 + 6 + 4	4 + 7 + 4	4 + 8 + 4
# 16	5 + 5 + 5	5 + 6 + 5	5 + 7 + 5	5 + 8 + 5
# 17	6 + 5 + 6	6 + 6 + 6	6 + 7 + 6	6 + 1 + 2
# 18	7 + 5 + 7	7 + 1 + 2	3 + 1 + 2	7 + 1 + 2
# 19	4 + 9 + 4	1 + 1 + 2	4 + 1 + 2	8 + 1 + 2
# 20	5 + 9 + 5	2 + 1 + 2	5 + 1 + 2	9 + 1 + 2

Grade 1 Chapter 3 Lesson 2

Addition three numbers in row 2

<u>Three numbers can add in any order</u>

9 + 5 + 3 = 14 + 3 = 17

9 + 5 + 3 = 9 + 8 = 17

9+3+5=12+5=17

9+3+5 = 9+8=17

Add numbers

	a	b	c	d
# 1	0 + 1 + 2	0 + 2 + 3	0 + 3 + 4	0 + 4 + 5
# 2	1 + 1 + 2	1 + 2 + 3	1 + 3 + 4	1 + 4 + 5
# 3	2 + 1 + 2	2 + 2 + 3	2 + 3 + 4	2 + 4 + 5
# 4	3 + 1 + 2	3 + 2 + 3	3 + 3 + 4	3 + 4 + 5
# 5	4 + 1 + 2	4 + 2 + 3	4 + 3 + 4	4 + 4 + 5
# 6	5 + 1 + 2	5 + 2 + 3	5 + 3 + 4	5 + 4 + 5
# 7	6 + 1 + 2	6 + 2 + 3	6 + 3 + 4	6 + 4 + 5
# 8	7 + 1 + 2	7 + 2 + 3	7 + 3 + 4	7 + 4 + 5
# 9	8 + 1 + 2	8 + 2 + 3	8 + 3 + 4	8 + 4 + 5
# 10	9 + 1 + 2	9 + 2 + 3	9 + 3 + 4	9 + 4 + 5
# 11	0 + 1 + 3	0 + 2 + 4	0 + 3 + 4	0 + 4 + 5
# 12	1 + 1 + 3	1 + 2 + 4	1 + 3 + 4	1 + 4 + 5
# 13	2 + 1 + 3	2 + 2 + 4	2 + 3 + 4	2 + 4 + 5
# 14	3 + 1 + 3	3 + 2 + 4	3 + 3 + 4	3 + 4 + 5
# 15	4 + 1 + 3	4 + 2 + 4	4 + 3 + 4	4 + 4 + 5
# 16	5 + 1 + 3	5 + 2 + 4	5 + 3 + 4	5 + 4 + 5
# 17	6 + 1 + 3	6 + 2 + 4	6 + 3 + 4	6 + 4 + 5
# 18	7 + 1 + 3	7 + 2 + 4	7 + 3 + 4	7 + 4 + 5
# 19	8 + 1 + 3	8 + 2 + 4	8 + 3 + 4	8 + 4 + 5
# 20	9 + 1 + 3	9 + 2 + 4	9 + 3 + 4	9 + 4 + 5

Grade 1 Chapter 3 Lesson 3

Subtraction three numbers in row 1

<u>To subtract numbers, work from left to right</u>

10 – 1 – 4 = 9 – 4 = 5

10 – 1 – 5 = 9 – 5 = 4

Subtract numbers

	a	b	c	d
# 1	1 - 1 - 0	1 - 0 - 1	2 - 1 - 1	4 – 1 - 2
# 2	2 - 1 - 0	2 - 0 - 1	3 - 1 – 1	5 – 1 - 2
# 3	3 - 1 - 0	3 - 0 - 1	4 - 1 – 1	6 – 1 - 2
# 4	4 - 1 - 0	4 - 0 - 1	5 - 1 – 1	7 – 1 - 2
# 5	5 - 1 - 0	5 - 0 - 1	6 - 1 – 1	8 – 1 - 2
# 6	6 - 1 - 0	6 - 0 - 1	7 - 1 – 1	9 – 1 - 2
# 7	7 - 1 - 0	7 - 0 - 1	8 - 1 – 1	10 – 1 - 2
# 8	8 - 1 - 0	8 - 0 - 1	9 - 1 – 1	11 – 1 - 2
# 9	9 - 1 - 0	9 - 0 - 1	10 - 1 – 1	12 – 1 - 2
# 10	10 - 1 - 0	10 - 0 - 1	11 - 1 – 1	13 – 1 - 2
# 11	4 - 2 - 1	5 - 1 - 3	5 - 3 - 1	5 – 2 - 2
# 12	5 - 2 - 1	6 - 1 - 3	6 - 3 – 1	6 – 2 - 2
# 13	6 - 2 - 1	7 - 1 - 3	7 - 3 - 1	7 – 2 - 2
# 14	7 - 2 - 1	8 - 1 - 3	8 - 3 – 1	8 – 2 - 2
# 15	8 - 2 - 1	9 - 1 - 3	9 - 3 - 1	9 – 2 - 2
# 16	9 - 2 - 1	10 - 1 - 3	10 - 3 – 1	10 – 2 - 2
# 17	10 - 2 - 1	11 - 1 - 3	11 - 3 – 1	11 – 2 - 2
# 18	11 - 2 - 1	12 - 1 - 3	12 - 3 – 1	12 – 2 - 2
# 19	12 - 2 - 1	13 - 1 - 3	13 - 3 – 1	13 – 2 - 2
# 20	13 - 2 - 1	14 - 1 - 3	14 - 3 – 1	14 – 2 - 2

Grade 1 Chapter 3 Lesson 4

Subtraction three numbers in row 2

To subtract numbers, work from left to right

10 – 1 – 6 = 9 – 6 = 3

10 – 1 – 7 = 9 – 7 = 2

Subtract numbers

	a	b	c	d
# 1	6 - 1 - 4	6 - 4 - 1	6 - 2 - 3	6 – 3 - 2
# 2	7 - 1 - 4	7 - 4 - 1	7 - 2 – 3	7 – 3 - 2
# 3	8 - 1 - 4	8 - 4 - 1	8 - 2 – 3	8 – 3 - 2
# 4	9 - 1 - 4	9 - 4 - 1	9 - 2 – 3	9 – 3 - 2
# 5	10 - 1 - 4	10 - 4 - 1	10 - 2 – 3	10 – 3 - 2
# 6	11 - 1 - 4	11 - 4 - 1	11 - 2 – 3	11 – 3 - 2
# 7	12 - 1 - 4	12 - 4 - 1	12 - 2 – 3	12 – 3 - 2
# 8	13 - 1 - 4	13 - 4 - 1	13 - 2 – 3	13 – 3 – 2
# 9	14 - 1 - 4	14 - 4 - 1	14 - 2 – 3	14 – 3 - 2
# 10	15 - 1 - 4	15 - 4 - 1	15 - 2 – 3	15 – 3 - 2
# 11	7 - 1 - 6	7 - 6 - 1	7 - 2 - 5	7 – 5 - 2
# 12	8 - 1 - 6	8 - 6 - 1	8 - 2 – 5	8 – 5 - 2
# 13	9 - 1 - 6	9 - 6 - 1	9 - 2 - 5	9 – 5 - 2
# 14	10 - 1 - 6	10 - 6 - 1	10 - 2– 5	10 – 5 - 2
# 15	11 - 1 - 6	11 - 6 - 1	11 - 2 - 5	11 – 5 - 2
# 16	12 - 1 - 6	12 - 6 – 1	12 - 2 – 5	12 – 5 - 2
# 17	13 - 1 - 6	13 - 6 - 1	13 - 2 – 5	13 – 5 - 2
# 18	14 - 1 - 6	14 - 6 - 1	14 - 2 – 5	14 – 5 - 2
# 19	15 - 1 - 6	15 - 6 - 1	15 - 2 – 5	15 – 5 - 2
# 20	16 - 1 - 6	16 - 6 - 1	16 - 2 – 5	16 – 5 - 2

Grade 1 Chapter 3 Lesson 5

Remember

10 = 9 + 1= 8 + 2= 7 + 3 = 6 + 4 = 5 + 5

11 = 9 + 2 = 8 + 3 = 7 + 4 = 6 + 5

12 = 9 +3 = 8 + 4 = 7 + 5 = 6 + 6 13 = 9 + 4 = 8 + 5 = 7 + 6

Two operations with three numbers in row 1

To add and subtract numbers, work from left to right

5 + 6 – 9 = 11 – 9 = 2 5 + 7 – 9 = 12 - 9 = 3

Add and Subtract numbers

	a	b	c	d
# 1	1 + 8 - 0	2 + 7 - 0	4 + 6 – 0	5 + 9 - 0
# 2	1 + 8 - 1	2 + 7 - 1	4 + 6 – 1	5 + 9 - 1
# 3	1 + 8 - 2	2 + 7 - 2	4 + 6 – 2	5 + 9 - 2
# 4	1 + 8 - 3	2 + 7 - 3	4 + 6 – 3	5 + 9 - 3
# 5	1 + 8 - 4	2 + 7 - 4	4 + 6 – 4	5 + 9 - 4
# 6	1 + 8 - 5	2 + 7 - 5	4 + 6 – 5	5 + 9 - 5
# 7	1 + 8 - 6	2 + 7 - 6	4 + 6 – 6	5 + 9 - 6
# 8	1 + 8 - 7	2 + 7 - 7	4 + 6 – 7	5 + 9 - 7
# 9	1 + 8 - 8	2 + 7 - 8	4 + 6 – 8	5 + 9 - 8
# 10	1 + 8 - 9	2 + 7 - 9	4 + 6 – 9	5 + 9 - 9
# 11	6 + 4 - 0	3 + 7 - 0	7 + 4 - 0	8 + 5 - 0
# 12	6 + 4 - 1	3 + 7 - 1	7 + 4 – 1	8 + 5 - 1
# 13	6 + 4 - 2	3 + 7 - 2	7 + 4 - 2	8 + 5 - 2
# 14	6 + 4 - 3	3 + 7 - 3	7 + 4 – 3	8 + 5 - 3
# 15	6 + 4 - 4	3 + 7 - 4	7 + 4 - 4	8 + 5 - 4
# 16	6 + 4 - 5	3 + 7 – 5	7 + 4 – 5	8 + 5 - 5
# 17	6 + 4 - 6	3 + 7 - 6	7 + 4 – 6	8 + 5 - 6
# 18	6 + 4 - 7	3 + 7 - 7	7 + 4 – 7	8 + 5 - 7
# 19	6 + 4 - 8	3 + 7 - 8	7 + 4 – 8	8 + 5 - 8
# 20	6 + 4 - 9	3 + 7 - 9	7 + 4 – 9	8 + 5 - 9

Grade 1 Chapter 3 Lesson 6

Remember

14 = 9 + 5 = 8 + 6 = 7 + 7　　15 = 9 + 6 = 8 + 7

16 = 9 + 7 = 8 + 8　　17 = 9 + 8

18 = 9 + 9　　19 = 10 + 9

Two operations with three numbers in row 2

<u>To subtract and add numbers, work from left to right</u>

9 - 8 + 7 = 1 + 7 = 8

8 – 7 + 6 = 1 + 6 = 7

Subtract and add numbers

	a	b	c	d
# 1	4 - 3 + 0	5 - 3 + 0	6 - 3 + 0	7 – 3 + 0
# 2	4 - 3 + 1	5 - 3 + 1	6 - 3 + 1	7 – 3 + 1
# 3	4 - 3 + 2	5 - 3 + 2	6 - 3 + 2	7 – 3 + 2
# 4	4 - 3 + 3	5 - 3 + 3	6 - 3 + 3	7 – 3 + 3
# 5	4 - 3 + 4	5 - 3 + 4	6 - 3 + 4	7 – 3 + 4
# 6	4 - 3 + 5	5 - 3 + 5	6 - 3 + 5	7 – 3 + 5
# 7	4 - 3 + 6	5 - 3 + 6	6 - 3 + 6	7 – 3 + 6
# 8	4 - 3 + 7	5 - 3 + 7	6 - 3 + 7	7 – 3 + 7
# 9	4 - 3 + 8	5 - 3 + 8	6 - 3 + 8	7 – 3 + 8
# 10	4 - 3 + 9	5 - 3 + 9	6 - 3 + 9	7 – 3 + 9
# 11	6 - 4 + 0	7 - 4 + 0	8 - 4 + 0	9 – 4 + 0
# 12	6 - 4 + 1	7 - 4 + 1	8 - 4 + 1	9 – 4 + 1
# 13	6 - 4 + 2	7 - 4 + 2	8 - 4 + 2	9 – 4 + 2
# 14	6 - 4 + 3	7 - 4 + 3	8 – 4 + 3	9 – 4 + 3
# 15	6 - 4 + 4	7 - 4 + 4	8 - 4 + 4	9 – 4 + 4
# 16	6 - 4 + 5	7 - 4 + 5	8 - 4 + 5	9 – 4 + 5
# 17	6 - 4 + 6	7 - 4 + 6	8 - 4 + 6	9 – 4 + 6
# 18	6 - 4 + 7	7 - 4 + 7	8 - 4 + 7	9 – 4 + 7
# 19	6 - 4 + 8	7 - 4 + 8	8 - 4 + 8	9 – 4 + 8
# 20	6 - 4 + 9	7 - 4 + 9	8 - 4 + 9	9 – 4 + 9

Grade 1 Chapter 3 Lesson 7

Addition tens in column

80 + 10 = 90 or		10 + 80 = 90	
80	1) 0 + 0 = 0	10	1) 0 + 0 =0
+ 10	2) Put 0 below 0	+80	2) Put 0 below 0
90	3) 8 + 1 = 9	90	3) 1 + 8 = 9
	4) Put 9 below 1		4) Put 9 below 8

Add tens

	a	b	c	d
# 1	10 + 10	10 + 20	10 + 30	10 + 40
# 2	10 + 50	10 + 60	10 + 70	10 + 80
# 3	20 + 10	20 + 20	20 + 30	20 + 40
# 4	20 + 50	20 + 60	20 + 70	70 + 20
# 5	30 +10	30 + 20	30 + 30	30 + 40
# 6	30 + 50	30 + 60	50 + 30	60 + 30
# 7	40 + 10	40 + 20	40 + 30	40 + 40
# 8	40 + 50	20 + 40	30 + 40	20 + 60
# 9	50 + 10	50 + 20	50 + 30	50 + 40
# 10	60 + 10	60 + 20	60 + 30	30 + 60

Grade 1 Chapter 3 Lesson 8

Subtraction tens in column

90 – 10 = 80 or

```
 90     1) 0 - 0 = 0
-10     2) Put 0 below 0
 80     3) 9 – 1 = 8
        4) Put 8 below 1
```

90 – 80 = 10

```
 90     1) 0 – 0 = 0
-80     2) Put 0 below 0
 10     3) 9 – 8 = 1
        4) Put 1 below 8
```

Subtract tens

	a	b	c	d
# 1	90 - 10	90 - 20	90 - 30	90 - 40
# 2	90 - 50	90 - 60	90 - 70	90 - 80
# 3	90 - 90	80 - 10	80 - 20	80 - 30
# 4	80 - 40	80 - 50	80 - 60	80 - 70
# 5	80 - 80	70 - 10	70 - 20	70 - 30
# 6	70 - 40	70 - 50	70 - 60	70 - 70
# 7	60 - 10	60 - 20	60 - 30	60 - 40
# 8	60 - 50	60 - 60	50 - 10	50 - 20
# 9	50 - 30	50 - 40	50 - 50	40 - 10
# 10	40 - 20	40 - 30	30 - 10	30 - 20

Grade 1 Chapter 3 Lesson 9
Addition two-digit numbers in column
Add the ones and then add the tens

43	1) 3 + 5 =8	27	1) 7 + 1 = 8
+ 25	2) Put 8 below 5	+ 51	2) Put 8 below 1
68	3) 4 + 2 = 6	78	3) 2 + 5 = 7
	4) Put 6 below 2		4) Put 7 below 5

Add numbers

	a	b	c	d
# 1	20 + 79	21 + 68	22 + 57	23 + 46
# 2	24 + 35	25 + 23	26 + 12	27 + 41
# 3	28 + 70	27 + 60	30 + 58	31 + 66
# 4	32 + 65	33 + 53	34 + 42	35 + 34
# 5	36 + 20	37 + 10	38 + 31	39 + 10
# 6	40 + 25	41 + 35	42 + 44	43 + 52
# 7	44 + 40	45 + 30	46 + 21	47 + 11
# 8	48 + 20	49 + 30	50 + 42	51 + 36
# 9	52 + 45	53 + 30	54 + 21	55 + 11
# 10	56 + 23	50 + 32	58 + 41	59 + 30

Grade 1 Chapter 3 Lesson 10

Subtraction two-digit numbers in column

Subtract the ones then and subtract the tens

96 - 25 71	1) 6 – 5 = 1 2) Put 1 below 5 3) 9 – 2 = 7 4) Put 7 below 2	89 - 52 37	1) 9 – 2 = 7 2) Put 7 below 2 3) 8 – 5 = 3 4) Put 3 below 5

Subtract numbers

	a	b	c	d
# 1	99 - 87	98 - 76	97 - 65	96 - 54
# 2	95 - 43	94 - 32	93 - 21	92 - 10
# 3	91 - 20	90 - 30	89 - 55	88 - 74
# 4	87 - 56	86 - 43	85 - 41	84 - 31
# 5	85 - 31	82 - 20	81 - 11	86 - 14
# 6	79 - 22	78 - 33	77 - 44	76 - 50
# 7	75 - 61	74 - 52	73 - 41	72 - 31
# 8	79 - 44	70 - 40	69 - 59	68 - 43
# 9	67 - 54	66 - 41	65 - 32	64 - 21
# 10	97 - 52	86 - 42	75 - 34	69 - 35

Grade 1 Chapter 3 Lesson 11
Addition three numbers in column
Add the ones and then add the tens

10	1) 0 + 1 + 2 = 3
+ 21	2) Put 3 below 2
32	3) 1 + 2 + 3 = 6
63	4) Put 6 below 3

Add numbers

	a	b	c	d
# 1	20	21	22	23
	+ 12	+ 13	+ 14	+ 15
	32	31	30	30
# 2	24	25	26	23
	+ 11	+ 12	+ 11	+ 10
	23	22	20	22
# 3	28	20	30	31
	+ 20	+ 21	+ 22	+ 23
	21	22	23	22
# 4	32	33	34	33
	+ 21	+ 25	+ 23	+ 20
	20	10	12	14
# 5	31	32	33	34
	+ 22	+ 24	+ 22	+ 21
	12	11	11	21

Grade 1 Chapter 3 Lesson 11

Addition three numbers in column

Add the ones and then add the tens

21	1) 1 + 2 + 3 = 6
+ 32	2) Put 6 below 3
43	3) 2 + 3 + 4 = 9
96	4) Put 9 below 4

Add numbers

	a	b	c	d
# 6	36	37	36	30
	+ 20	+ 20	+ 20	+ 31
	12	12	11	13
# 7	40	41	42	43
	+ 32	+ 33	+ 34	+ 35
	12	11	10	11
# 8	44	45	46	44
	+ 32	+ 30	+ 32	+ 31
	12	11	11	10
# 9	48	43	50	31
	+ 40	+ 40	+ 32	+ 43
	11	10	12	12
# 10	21	34	24	35
	+ 31	+ 22	+ 32	+ 21
	24	21	22	23

Grade 1 Chapter 3 Lesson 12
Add and Subtract numbers
Test 3 A
Add and Subtract numbers in row

 # 1 9 + 6 + 4
 # 2 8 + 6 + 5
 # 3 10 – 3 - 1
 # 4 10 – 4 - 2
 # 5 6 + 6 - 2

 # 6 6 + 7 - 3
 # 7 10 - 4 + 3
 # 8 10 – 5 + 4
 # 9 60 + 20
 # 10 60 - 20

Add and Subtract numbers in column

# 11	# 12	# 13	# 14	# 15
10	90	51	95	35
+ 80	- 70	+ 48	- 41	+ 23
				11

# 16	# 17	# 18	# 19	# 20
20	80	62	84	24
+ 60	- 50	+ 36	- 31	+ 21
				32

Grade 1 Chapter 3 Lesson 12
Add and Subtract numbers
Test 3 B
Add and Subtract Numbers in row

1 7 + 5 + 4
2 6 + 4 + 3
3 10 – 5 - 3
4 10 – 6 – 4
5 6 + 8 - 4

6 6 + 9 - 5
7 10 – 6 + 5
8 10 – 7 + 6
9 40 + 20
10 40 – 20

Add and Subtract Numbers in column

# 11 30	# 12 70	# 13 73	# 14 74	# 15 42
+ 40	- 30	+ 24	- 21	+ 20
				34

# 16 60	# 17 60	# 18 84	# 19 62	# 20 51
+ 20	- 10	+12	- 11	+ 13
				23

Grade 1 Answers Chapter 3

Lesson 1

	a	b	c	d
# 1	1	2	3	4
# 2	3	4	5	6
# 3	5	6	7	8
# 4	7	8	9	10
# 5	9	10	11	12
# 6	11	12	13	14
# 7	13	14	15	16
# 8	15	16	17	18
# 9	17	18	19	13
# 10	19	9	11	15

Lesson 1

	a	b	c	d
# 11	5	6	7	8
# 12	7	8	9	10
# 13	9	10	11	12
# 14	11	12	13	14
# 15	13	14	15	16
# 16	15	16	17	18
# 17	17	18	19	9
# 18	19	10	6	10
# 19	17	4	7	11
# 20	19	5	8	12

Lesson 2

	a	b	c	d
# 1	3	5	7	9
# 2	4	6	8	10
# 3	5	7	9	11
# 4	6	8	10	12
# 5	7	9	11	13
# 6	8	10	12	14
# 7	9	11	13	15
# 8	10	12	14	16
# 9	11	13	15	17
# 10	12	14	16	18

Grade 1 Answers Chapter 3

Lesson 2

# 11	4	6	7	9
# 12	5	7	8	10
# 13	6	8	9	11
# 14	7	9	10	12
# 15	8	10	11	13
# 16	9	11	12	14
# 17	10	12	13	15
# 18	11	13	14	16
# 19	12	14	15	17
# 20	13	15	16	18

Lesson 3

	a	b	c	d
# 1	0	0	0	1
# 2	1	1	1	2
# 3	2	2	2	3
# 4	3	3	3	4
# 5	4	4	4	5
# 6	5	5	5	6
# 7	6	6	6	7
# 8	7	7	7	8
# 9	8	8	8	9
# 10	9	9	9	10

Lesson 3

	a	b	c	d
# 11	1	1	1	1
# 12	2	2	2	2
# 13	3	3	3	3
# 14	4	4	4	4
# 15	5	5	5	5
# 16	6	6	6	6
# 17	7	7	7	7
# 18	8	8	8	8
# 19	9	9	9	9
# 20	10	10	10	10

Grade 1 Answers Chapter 3

Lesson 4

	a	b	c	d
# 1	1	1	1	1
# 2	2	2	2	2
# 3	3	3	3	3
# 4	4	4	4	4
# 5	5	5	5	5
# 6	6	6	6	6
# 7	7	7	7	7
# 8	8	8	8	8
# 9	9	9	9	9
# 10	10	10	10	10

Lesson 4

	a	b	c	d
# 11	0	0	0	0
# 12	1	1	1	1
# 13	2	2	2	2
# 14	3	3	3	3
# 15	4	4	4	4
# 16	5	5	5	5
# 17	6	6	6	6
# 18	7	7	7	7
# 19	8	8	8	8
# 20	9	9	9	9

Lesson 5

	a	b	c	d
# 1	9	9	10	14
# 2	8	8	9	13
# 3	7	7	8	12
# 4	6	6	7	11
# 5	5	5	6	10
# 6	4	4	5	9
# 7	3	3	4	8
# 8	2	2	3	7
# 9	1	1	2	6
# 10	0	0	1	5

Lesson 5

	a	b	c	d
# 11	10	10	11	13
# 12	9	9	10	12
# 13	8	8	9	11
# 14	7	7	8	10
# 15	6	6	7	9
# 16	5	5	6	8
# 17	4	4	5	7
# 18	3	3	4	6
# 19	2	2	3	5
# 20	1	1	2	4

Lesson 6

	a	b	c	d
# 1	1	2	3	4
# 2	2	3	4	5
# 3	3	4	5	6
# 4	4	5	6	7
# 5	5	6	7	8
# 6	6	7	8	9
# 7	7	8	9	10
# 8	8	9	10	11
# 9	9	10	11	12
# 10	10	11	12	13

Lesson 6

	a	b	c	d
# 11	2	3	4	5
# 12	3	4	5	6
# 13	4	5	6	7
# 14	5	6	7	8
# 15	6	7	8	9
# 16	7	8	9	10
# 17	8	9	10	11
# 18	9	10	11	12
# 19	10	11	12	13
# 20	11	12	13	14

Grade 1 Answers Chapter 3

Lesson 7

	a	b	c	d
# 1	20	30	40	50
# 2	60	70	80	90
# 3	30	40	50	60
# 4	70	80	90	90
# 5	40	50	60	70
# 6	80	90	80	90
# 7	50	60	70	80
# 8	90	60	70	80
# 9	60	70	80	90
# 10	70	80	90	90

Lesson 8

	a	b	c	d
# 1	80	70	60	50
# 2	40	30	20	10
# 3	0	70	60	50
# 4	40	30	20	10
# 5	0	60	50	40
# 6	30	20	10	0
# 7	50	40	30	20
# 8	10	0	40	30
# 9	20	10	0	30
# 10	20	10	20	10

Lesson 9

	a	b	c	d
# 1	99	89	79	69
# 2	59	48	38	68
# 3	98	87	88	97
# 4	97	86	76	69
# 5	56	47	69	49
# 6	65	76	86	95
# 7	84	75	67	58
# 8	68	79	92	87
# 9	97	83	75	66
# 10	79	82	99	89

Lesson 10

	a	b	c	d
# 1	12	22	32	42
# 2	52	62	72	82
# 3	71	60	34	14
# 4	31	43	44	53
# 5	54	62	70	72
# 6	57	45	33	26
# 7	14	22	32	41
# 8	35	30	10	25
# 9	13	25	33	43
# 10	45	44	41	34

Lesson 11

	a	b	c	d
# 1	64	65	66	68
# 2	58	59	57	55
# 3	69	63	75	76
# 4	73	68	69	67
# 5	65	67	66	76
# 6	68	69	67	74
# 7	84	85	86	89
# 8	88	86	89	85
# 9	99	93	94	86
#10	76	77	78	79

Lesson 12

Test 3 A

# 1	19	# 2	19	# 3	6	# 4	4	# 5	10
# 6	10	# 7	9	# 8	9	# 9	80	# 10	40
# 11	90	# 12	20	# 13	99	# 14	54	# 15	69
# 16	80	# 17	30	# 18	98	# 19	53	# 20	77

Test 3 B

# 1	16	# 2	13	# 3	2	# 4	0	# 5	10
# 6	10	# 7	9	# 8	9	# 9	60	# 10	20
# 11	70	# 12	40	# 13	97	# 14	53	# 15	96
# 16	80	# 17	50	# 18	96	# 19	51	# 20	87

Grade 1
Chapter 4

Finding unknowns

(0 – 19)

Grade 1 Chapter 4 Contents
Finding unknowns (0-19)
Lesson 1 - 12
Advanced Problems

Work with lesson 1, read rules and examples, 2) Solve the problems from contents, 3) Check answers in the end of chapter 1, 4) Work with the next lessons 2, 3, 4, 5, …, 5) Solve Test 1 A, 6) Grading Scale: A =10 points, B=9points, C=8 points, D=7 points, F=0-6 points, 7) If you solve 10 problems correctly, you get an A, 8) If have an A, you can move on to next chapter, 9) If your grade is lower than an A, work with the mistakes, then solve Test 1B, 10) When you finish all of the chapters, solve Final Exam in last chapter.

Grade 1 Chapter 4 Lesson 1
Addition 1

Addition 2 + 3 = 5 5 > 2 5 > 3

Addend + Addend = Sum

First Addend = Sum – Second Addend ? + 3 = 5 5 – 3 = 2

Second Addend = Sum – First Addend 2 + ? = 5 5 – 2 = 3

Put correct number instead the question mark

	a	b	c	d
# 1	? + 0 = 0	? + 0 = 1	? + 0 = 2	? + 0 = 3
# 2	0 + ? = 1	1 + ? = 2	2 + ? = 3	3 + ? = 4
# 3	0 + 2 = ?	1 + 2 = ?	2 + 2 = ?	3 + 2 = ?
# 4	? + 3 = 3	? + 3 = 4	? + 3 = 5	? + 3 = 6
# 5	0 + ? = 4	1 + ? = 5	2 + ? = 6	3 + ? = 7
# 6	0 + 5 = ?	1 + 5 = ?	2 + 5 = ?	3 + 5 = ?
# 7	? + 6 = 6	? + 6 = 7	? + 6 = 8	? + 6 = 9
# 8	0 + ? = 7	1 + ? = 8	2 + ? = 9	3 + ? = 10
# 9	0 + 8 = ?	1 + 8 = ?	2 + 8 = ?	3 + 8 = ?
# 10	? + 9 = 9	? + 9 =10	? + 9 = 11	? + 9 = 12
# 11	? + 0 = 4	? + 0 = 5	? + 0 = 6	? + 0= 7
# 12	4 + ? = 5	5 + ? = 6	6 + ? = 7	7 + ? = 8
# 13	4 + 2 = ?	5 + 2 = ?	6 + 2 = ?	7 + 2 = ?
# 14	? + 3 = 7	? + 3 = 8	? + 3 = 9	? + 3 = 10
# 15	4 + ? = 8	5 + ? = 9	6 + ? = 10	7 + ? = 11
# 16	4 + 5 = ?	5 + 5 = ?	6 + 5 = ?	7 + 5 = ?
# 17	? + 6 = 10	? + 6 = 11	? + 6 = 12	? + 6 = 13
# 18	4 + ? = 11	5 + ? = 12	6 + ? = 13	7 + ? = 14
# 19	4 + 8 = ?	5 + 8 = ?	6 + 8 = ?	7 + 8 = ?
# 20	? + 9 = 13	? + 9 = 14	? + 9 = 15	? + 9 = 16

Grade 1 Chapter 4 Lesson 2
Addition 2

Addition $2 + 3 = 5$ $5 > 2$ $5 > 3$

Addend + Addend = Sum

First Addend = Sum – Second Addend	$? + 3 = 5$	$5 - 3 = 2$
Second Addend = Sum – First Addend	$2 + ? = 5$	$5 - 2 = 3$

Property of addition: Two numbers can add in any order.
The sum will be the same. $2 + 3 = 5$ $3 + 2 = 5$

Put correct number instead the question mark

	a	b	c	d
# 1	? + 0 = 8	? + 0 = 9	? + 0 = 10	? + 0 = 11
# 2	8 + ? = 9	9 + ? = 10	10 + ? = 11	11 + ? = 12
# 3	8 + 2 = ?	9 + 2 = ?	10 + 2 = ?	11 + 2 = ?
# 4	? + 3 = 11	? + 3 = 12	? + 3 = 13	? + 3 = 14
# 5	8 + ? = 12	9 + ? = 13	10 + ? = 14	11 + ? = 15
# 6	8 + 5 = ?	9 + 5 = ?	10 + 5 = ?	11 + 5 = ?
# 7	? + 6 = 14	? + 6 = 15	? + 6 = 16	? + 6 = 17
# 8	8 + ? = 15	9 + ? = 16	10 + ? = 17	11 + ? = 18
# 9	8 + 8 = ?	9 + 8 = ?	10 + 8 = ?	11 + 8 = ?
# 10	? + 9 = 17	? + 9 = 18	? + 9 = 19	? + 11 = 19

	a	b
# 11	? + 10 = 10 + 0	? + 11 = 11 + 0
# 12	1 + ? = 10 + 1	1 + ? = 11 + 1
# 13	2 + 10 = ? + 2	2 + 11 = ? + 2
# 14	3 + 10 = 10 + ?	3 + 11 = 11 + ?
# 15	? + 10 = 10 + 4	? + 11 = 11 + 4
# 16	5 + ? = 10 + 5	5 + ? = 11 + 5
# 17	6 + 10 = ? + 6	6 + 11 = ? + 6
# 18	7 + 10 = 10 + ?	7 + 11 = 11 + ?
# 19	? + 10 = 10 + 8	? + 11 = 11 + 8
# 20	9 + 10 = 10 + ?	11 + 9 = ? + 11

Grade 1 Chapter 4 Lesson 3
Subtraction 1

Subtraction 5 – 2 = 3 **5 > 2 5 > 3**

Minuend – Subtrahend = Difference

Minuend = Subtrahend + Difference **<u>? – 2 = 3 3 + 2 = 5</u>**

Subtrahend = Minuend – Difference **<u>5 - ? = 3 5 – 3 = 2</u>**

Put correct number instead the question mark

	a	b	c	d
# 1	? – 0 = 8	? – 0 = 9	? – 0 = 10	? – 0 = 11
# 2	8 – ? = 7	9 – ? = 8	10 – ? = 9	11 - ? = 10
# 3	? – 2 = 6	? – 2 = 7	? – 2 = 8	? – 2 = 9
# 4	8 – ? = 5	9 – ? = 6	10 – ? = 7	11 - ? = 8
# 5	? – 4 = 4	? – 4 = 5	? – 4 = 6	? – 4 = 7
# 6	8 – ? = 3	9 – ? = 4	10 – ? = 5	11 - ? = 6
# 7	? – 6 = 2	? – 6 = 3	? – 6 = 4	? – 6 = 5
# 8	8 – ? = 1	9 – ? = 2	10 – ? = 3	11 - ? = 4
# 9	? – 8 = 0	? – 8 = 1	? – 8 = 2	? – 8 = 3
# 10	7 – ? = 0	9 – ? = 0	10 – ? = 1	11 - ? = 1
# 11	? – 0 = 12	? – 0 = 13	? – 0 = 14	? – 0 = 15
# 12	12 – ? = 11	13 – ? = 12	14 – ? = 13	15 - ? = 14
# 13	? – 2 = 10	? – 2 = 11	? – 2 = 12	? – 2 = 13
# 14	12 – ? = 9	13 – ? = 10	14 – ? = 11	15 - ? = 12
# 15	? – 4 = 8	? – 4 = 9	? – 4 = 10	? – 4 = 11
# 16	12 – ? = 7	13 – ? = 8	14 – ? = 9	15 - ? = 10
# 17	? – 6 = 6	? – 6 = 7	? – 6 = 8	? – 6 = 9
# 18	12 – ? = 5	13 – ? = 6	14 – ? = 7	15 - ? = 8
# 19	? – 8 = 4	? – 8 = 5	? – 8 = 6	? – 8 = 7
# 20	12 – ? = 3	13 – ? = 4	14 – ? = 5	15 - ? = 6

Grade 1 Chapter 4 Lesson 4
Subtraction 2

Subtraction 5 – 2 = 3 **5 > 2 5 > 3**

Minuend – Subtrahend = Difference

Minuend = Subtrahend + Difference **<u>? – 2 = 3 3 + 2 = 5</u>**

Subtrahend = Minuend – Difference **<u>5 - ? = 3 5 – 3 = 2</u>**

Put correct number instead the question mark

	a	b	c	d
# 1	? – 0 = 16	? – 0 = 17	? – 0 = 18	? – 0 = 19
# 2	16 – ? = 15	17 – ? = 16	18 – ? = 17	19 - ? = 18
# 3	? – 2 = 14	? – 2 = 15	? – 2 = 16	? – 2 = 17
# 4	16 – ? = 13	17 – ? = 14	18 – ? = 15	19 - ? = 16
# 5	? – 4 = 12	? – 4 = 13	? – 4 = 14	? – 4 = 15
# 6	16 – ? = 11	17 – ? = 12	18 – ? = 13	19 - ? = 14
# 7	? – 6 = 10	? – 6 = 11	? – 6 = 12	? – 6 = 13
# 8	16 – ? = 9	17 – ? = 10	18 – ? = 11	19 - ? = 12
# 9	? – 8 = 8	? – 8 = 9	? – 8 = 10	? – 8 = 11
# 10	16 – ? = 7	17 – ? = 8	18 – ? = 9	19 - ? = 10
# 11	16 – ? = 16	17 – ? = 17	18 –? =18	19 - ? = 19
# 12	? – 1 = 15	? – 1 = 16	? – 1 = 17	? – 1 = 18
# 13	16 – ? = 14	17 – ? = 15	18 –? = 16	19 - ? = 17
# 14	? – 3 = 13	? – 3 = 14	? – 3 = 15	? – 3 = 16
# 15	16 – ? = 12	17 – ? = 13	18 – ? =14	19 - ? = 15
# 16	? – 5 = 11	? – 5 = 12	? – 5 = 13	? – 5 = 14
# 17	16 – ? = 10	17 – ? = 11	18 – ? = 12	19 - ? = 13
# 18	? – 7 = 9	? – 7 = 10	? – 7 = 11	? – 7 = 12
# 19	16 – ? = 8	17 – ? = 9	18 – ? = 10	19 - ? = 11
# 20	? – 9 = 7	? – 9 = 8	? – 9 = 9	? – 9 = 10

Grade 1 Chapter 4 Lesson 5
Addition 3

Rules	Examples
Addition	$2 + 3 = 5$
Addend + Addend = Sum	$5 > 2$ $5 > 3$
First Addend = Sum – Second Addend	$? + 3 = 5$ $5 - 3 = 2$
Second Addend = Sum – First Addend	$2 + ? = 5$ $5 - 2 = 3$

Complete the tables

1

	a	b	c	d	e
Addend		5	11		12
Addend	0		0	5	
Sum	10	10		11	12

2

Addend	7		8	14	
Addend	5	0		0	5
Sum		13	13		14

3

	a				
Addend	?	4	10		11
Addend	1		1	6	
Sum	10	10		11	12

a) $? + 1 = 10$

4

			c		
Addend	6		7	13	
Addend	6	1	?	1	6
Sum		13	13		14

c) $7 + ? = 13$

5

Addend	?	3	9	?	10
Addend	2	?	2	7	?
Sum	10	10	?	11	12

Grade 1 Chapter 4 Lesson 5

Addition 3

Rules	Examples
Addition	2 + 3 = 5
Addend + Addend = Sum	5 > 2 5 > 3
First Addend = Sum – Second Addend	? + 3 = 5 5 – 3 = 2
Second Addend = Sum – First Addend	2 + ? = 5 5 – 2 = 3

Complete the tables

6

	a	b	c	d	e
Addend	5		6	12	
Addend	7	2		2	7
Sum		13	13		14

7

Addend		2	8		9
Addend	3		3	8	
Sum	10	10		11	12

8

			c		
Addend	4		5	11	
Addend	8	3	?	3	8
Sum		13	13		14

c) 5 + ? = 13

9

				d	
Addend		1	7	?	8
Addend	4		4	9	
Sum	10	10		11	12

d) ? + 9 = 11

10

Addend	3	?	4	10	?
Addend	9	4	?	4	9
Sum	?	13	13	?	14

Grade 1 Chapter 4 Lesson 5

Addition 3

Rules	Examples
Addition	**2 + 3 = 5**
Addend + Addend = Sum	**5 > 2 5 > 3**
First Addend = Sum – Second Addend	**? + 3 = 5 5 – 3 = 2**
Second Addend = Sum – First Addend	**2 + ? = 5 5 – 2 = 3**

Complete the tables

 # 11

	a	b	c	d	e
Addend		10	16		17
Addend	0		0	5	
Sum	15	15		16	17

 # 12

Addend	12		13	19	
Addend	5	0		0	5
Sum		18	18		19

 # 13

				d	
Addend		9	15	?	16
Addend	1		1	6	
Sum	15	15		16	17

d) ? + 6 = 16

 # 14

			c		
Addend	11		12	18	
Addend	6	1	?	1	6
Sum		18	18		19

c) 12 + ? = 18

 # 15

Addend	?	8	14	?	15
Addend	2	?	2	7	?
Sum	15	15	?	16	17

Grade 1 Chapter 4 Lesson 5
Addition 3

Rules	Examples
Addition	**2 + 3 = 5**
Addend + Addend = Sum	**5 > 2 5 > 3**
First Addend = Sum – Second Addend	**? + 3 = 5 5 – 3 = 2**
Second Addend = Sum – First Addend	**2 + ? = 5 5 – 2 = 3**

Complete the tables

# 16	a	b	c	d	e
Addend	10		11	17	
Addend	7	2		2	7
Sum		18	18		19

# 17					e
Addend		7	13		14
Addend	3		3	8	
Sum	15	15		16	17

# 18					e
Addend	9		10	16	?
Addend	8	3		3	8
Sum		18	18		19

e) ? + 8 = 19

# 19					e
Addend		6	12		13
Addend	4		4	9	?
Sum	15	15		16	17

e) 13 + ? = 17

# 20	a	b	c	d	e
Addend	8	?	9	15	?
Addend	9	4	?	4	9
Sum	?	18	18	?	19

Grade 1 Chapter 4 Lesson 6
Subtraction 3

Rules	Examples
Subtraction	$5 - 2 = 3$
Minuend – Subtrahend = Difference	$5 > 2$ $5 > 3$
Minuend = Subtrahend + Difference	$? - 2 = 3$ $3 + 2 = 5$
Subtrahend = Minuend – Difference	$5 - ? = 3$ $5 - 3 = 2$

Complete the tables

 # 1

	a	b	c	d	e
Minuend		10	11		12
Subtrahend	0		0	5	
Difference	10	5		6	12

 # 2

Minuend	12		13	14	
Subtrahend	5	0		0	5
Difference		13	8		9

 # 3

				d	
Minuend		10	11	?	12
Subtrahend	1		1	6	
Difference	9	4		5	11

d) $? - 6 = 5$

 # 4

			c		
Minuend	12		13	14	
Subtrahend	6	1	?	1	6
Difference		12	7		8

c) $13 - ? = 7$

 # 5

Minuend	?	10	11	?	12
Subtrahend	2	?	2	7	?
Difference	8	3	?	4	10

Grade 1 Chapter 4 Lesson 6
Subtraction 3

Rules — **Examples**

Subtraction — **5 – 2 = 3**

Minuend – Subtrahend = Difference — **5 > 2   5 > 3**

Minuend = Subtrahend + Difference — **? - 2 = 3   3 + 2 = 5**

Subtrahend = Minuend – Difference — **5 - ? = 3   5 – 3 = 2**

Complete the tables

 # 6

	a	b	c	d	e
Minuend	12		13	14	
Subtrahend	7	2		2	7
Difference		11	6		7

 # 7

Minuend		10	11		12
Subtrahend	3		3	8	
Difference	7	2		3	9

 # 8

					e
Minuend	12		13	14	?
Subtrahend	8	3		3	8
Difference		10	5		6

e) ? – 8 = 6

 # 9

					e
Minuend		10	11		12
Subtrahend	4		4	9	
Difference	6	1		2	8

e) 12 - ? = 8

 # 10

Minuend	12	?	13	14	?
Subtrahend	9	4	?	4	9
Difference	?	9	4	?	5

Grade 1 Chapter 4 Lesson 6
Subtraction 3

Rules | **Examples**

Subtraction | **5 – 2 = 3**

Minuend – Subtrahend = Difference | **5 > 2   5 > 3**

Minuend = Subtrahend + Difference | **? - 2 = 3   3 + 2 = 5**

Subtrahend = Minuend – Difference | **5 - ? = 3   5 – 3 = 2**

Complete the tables

 # 11

	a	b	c	d	e
Minuend		15	16		17
Subtrahend	0		0	5	
Difference	15	10		11	17

 # 12

Minuend	17		18	19	
Subtrahend	5	0		0	5
Difference		18	13		14

 # 13

b) 15 - ? = 9

		b			
Minuend		15	16		17
Subtrahend	1	?	1	6	
Difference	14	9		10	16

 # 14

e) ? – 6 = 13

					e
Minuend	17		18	19	?
Subtrahend	6	1		1	6
Difference		17	12		13

 # 15

Minuend	?	15	16	?	17
Subtrahend	2	?	2	7	?
Difference	13	8	?	9	15

Grade 1 Chapter 4 Lesson 6
Subtraction 3

Rules | **Examples**

Subtraction — $5 - 2 = 3$

Minuend – Subtrahend = Difference — $5 > 2 \quad 5 > 3$

Minuend = Subtrahend + Difference — $? - 2 = 3 \quad 3 + 2 = 5$

Subtrahend = Minuend – Difference — $5 - ? = 3 \quad 5 - 3 = 2$

Complete the tables

 # 16

	a	b	c	d	e
Minuend		15	16		17
Subtrahend	2		2	7	
Difference	13	8		9	15

 # 17

					e
Minuend		15	16		17
Subtrahend	3		3	8	
Difference	12	7		8	14

 # 18 — e) ? – 8 = 11

					e
Minuend	17		18	19	?
Subtrahend	8	3		3	8
Difference		15	10		11

 # 19 — e) 17 - ? = 13

					e
Minuend		15	16		17
Subtrahend	4		4	9	?
Difference	11	6		7	13

 # 20

	a	b	c	d	e
Minuend	17	?	18	19	?
Subtrahend	9	4	?	4	9
Difference	?	14	9	?	10

Grade 1 Chapter 4 Lesson 7
Review 4 A

Put correct number instead the question mark

\# 1 ? + 9 = 15

\# 2 8 + ? = 16

\# 3 ? + 7 = 17

\# 4 6 + ? = 18

\# 5 ? + 5 = 19

\# 6 ? - 5 = 14

\# 7 19 - ? = 13

\# 8 ? – 7 = 11

\# 9 18 - ? = 11

\# 10 ? – 9 = 10

Complete the tables

	# 11	# 12	# 13	# 14	# 15
Addend	?	8	9	?	12
Addend	4	?	6	7	?
Sum	15	16	?	18	19

	# 16	# 17	# 18	# 19	# 20
Minuend	?	16	17	?	19
Subtrahend	9	?	9	9	?
Difference	6	7	?	9	10

Grade 1 Chapter 4 Lesson 7
Review 4 B

Put correct number instead the question mark

 # 1 ? + 9 = 16

 # 2 9 + ? = 15

 # 3 ? + 9 = 14

 # 4 9 + ? = 13

 # 5 ? + 9 = 12

 # 6 ? - 7 = 11

 # 7 16 - ? = 10

 # 8 ? – 7 = 9

 # 9 17 - ? = 12

 # 10 ? – 6 = 11

Complete the tables

	# 11	# 12	# 13	# 14	# 15
Addend	?	5	5	?	7
Addend	7	?	8	6	?
Sum	11	12	?	13	14

	# 16	# 17	# 18	# 19	# 20
Minuend	?	14	15	?	19
Subtrahend	4	?	9	5	?
Difference	8	7	?	11	9

Grade 1 Chapter 4 Lesson 8
Test 4 A
Put right number instead question mark

1 ? + 7 = 10

2 5 + ? = 11

3 ? + 5 = 12

4 5 + ? = 13

5 ? + 8 = 14

6 ? - 8 = 2

7 11 - ? = 2

8 ? – 8 = 4

9 13 - ? = 9

10 ? – 8 = 6

Complete the tables

	# 11	# 12	# 13	# 14	# 15
Addend	?	9	9	?	11
Addend	6	?	8	8	?
Sum	15	16	?	18	19

	# 16	# 17	# 18	# 19	# 20
Minuend	?	16	17	?	19
Subtrahend	8	?	8	8	?
Difference	7	8	?	10	9

Grade 1 Chapter 4 Lesson 8
Test 4 B
Put right number instead question mark

1 ? + 8 = 10

2 4 + ? = 11

3 ? + 4 = 12

4 4 + ? = 13

5 ? + 9 = 14

6 ? - 7 = 3

7 11 - ? = 3

8 ? – 9 = 3

9 13 - ? = 8

10 ? – 9 = 5

Complete the tables

	# 11	# 12	# 13	# 14	# 15
Addend	?	8	9	?	9
Addend	8	?	8	9	?
Sum	15	16	?	18	19

	# 16	# 17	# 18	# 19	# 20
Minuend	?	16	17	?	19
Subtrahend	9	?	9	9	?
Difference	6	9	?	9	8

Grade 1 Chapter 4 Lesson 9
Summary 1
Memorize examples, its easy, and you improve your memory

Addition

Rules	Examples
Addition	$2 + 3 = 5$
Addend + Addend = Sum	$5 > 2 \quad 5 > 3$
First Addend = Sum – Second Addend	$\underline{? + 3 = 5 \quad 5 - 3 = 2}$
Second Addend = Sum – First Addend	$\underline{2 + ? = 5 \quad 5 - 2 = 3}$

Subtraction

Rules	Examples
Subtraction	$5 - 2 = 3$
Minuend – Subtrahend = Difference	$5 > 2 \quad 5 > 3$
Minuend = Subtrahend + Difference	$\underline{? - 2 = 3 \quad 3 + 2 = 5}$
Subtrahend = Minuend – Difference	$\underline{5 - ? = 3 \quad 5 - 3 = 2}$

Put correct number instead the question mark

	a	b	c	d
# 1	8 + 8 = ?	9 + 8 = ?	10 + 8 = ?	11 + 8 = ?
# 2	? + 9 = 17	? + 9 = 18	? + 9 = 19	? + 11 = 19
# 3	? – 7 = 9	? – 7 = 10	? – 7 = 11	? – 7 = 12
# 4	16 – ? = 8	17 – ? = 9	18 – ? = 10	19 - ? = 11
# 5	? – 9 = 7	? – 9 = 8	? – 9 = 9	? – 9 = 10

Grade 1 Chapter 4 Lesson 9
Summary 1

Complete the tables

 # 6

	a	b	c	d	e
Addend	?	6	12	?	13
Addend	4	?	4	9	?
Sum	15	15	?	16	17

 # 7

	a	b	c	d	e
Addend	8	?	9	15	?
Addend	9	4	?	4	9
Sum	?	18	18	?	19

c) 9 + ? = 18

 # 8

	a	b	c	d	e
Minuend	17	?	18	19	?
Subtrahend	8	3	?	3	8
Difference	?	15	10	?	11

b) ? – 3 = 15

 # 9

	a	b	c	d	e
Minuend	?	15	16	?	17
Subtrahend	4	?	4	9	?
Difference	11	6	?	7	13

e) 17 - ? = 13

 # 10

Minuend	17	?	18	19	?
Subtrahend	9	4	?	4	9
Difference	?	14	9	?	10

Grade 1 Chapter 4 Lesson 10
Summary 2 (Basic Problems)
Numbers from 0 to 9

Count forward from 1 to 9 1, 2, 3, 4, 5, 6, 7, 8, 9
One, two, three, four, five, six, seven, eight, nine
First, second, third, fourth, fifth, sixth, seventh, eighth, ninth
Count backwards from 9 to 1 9, 8, 7, 6, 5, 4, 3, 2, 1
The smallest number is 0 The biggest number is 9
Equal means $5 = 5$ Number zero is less than one $0 < 1$
Less than (<) means

$0 < 1$	$1 < 2$	$2 < 3$	$3 < 4$	$4 < 5$
$5 < 6$	$6 < 7$	$7 < 8$	$8 < 9$	$9 = 9$

More than (>) means

$9 > 8$	$8 > 7$	$7 > 6$	$6 > 5$	$5 > 4$
$4 > 3$	$3 > 2$	$2 > 1$	$1 > 0$	$0 = 0$

Addition numbers in row

The first operation with number is addition +
To add numbers put numbers together: $2 + 3 = 5$
How many in all or how many together? If put together 2 and 3 result is 5
If you need to add $5 + 3$ you have to count after 5 forward 3 times
That should be 6, 7, 8 that is why $5 + 3 = 8$
The sum of a number and zero is the same as the number
$5 + 0 = 5$ $0 + 5 = 5$
Property of addition. Two numbers can add in any order.
The sum wills the same. $2+3=5$ $3+2=5$
If you need to add $3 + 6$, easy to start from bigger number. You need to start after 6 by count forward 3 times, that should be 7, 8, 9
That is why $6+3=9$.

Subtract numbers in row

Second operation with numbers is subtraction –
To subtract numbers, take away something: $5 - 2 = 3$. How many are left? 3 are left. If you need to subtract $5 - 3$ you have to count backward from 5 three times 4, 3, 2, that why $5 - 3 = 2$
Any number subtract zero result is the same number $5 - 0 = 5$
Any number subtract the same number result is zero $5 - 5 = 0$

Grade 1 Chapter 4 Lesson 10
Summary 3 (Basic Problems)
Remember

10 = 9 + 1 = 8 + 2 = 7 + 3 = 6 + 4 = 5 + 5
11 = 9 + 2 = 8 + 3 = 7 + 4 = 6 + 5
12 = 9 + 3 = 8 + 4 = 7 + 5 = 6 + 6
13 = 9 + 4 = 8 + 5 = 7 + 6
14 = 9 + 5 = 8 + 6 = 7 + 7
15 = 9 + 6 = 8 + 7
16 = 9 + 7 = 8 + 8
17 = 9 + 8
18 = 9 + 9
19 = 10 + 9

One-digit numbers
One-digit numbers have one digit
This is numbers from 0 to 9: 0, 1, 2, 3, 4, 5, 6, 7, 8, 9

Two-digit numbers
Two-digit numbers have two digits. This is numbers from 10 to 99
Digit on the left are called tens. Digit on the right are called ones
Digits 0, 1, 2, 3, 4, 5, 6, 7, 8, 9
Read numbers 10 to 19: 10, 11, 12, 13, 14, 15, 16, 17, 18, 19
Read numbers 20 through 29: 20, 21, 22, 23, 24, 25, 26, 27, 28, 29
Read numbers 30 through 39: 30, 31, 32, 33, 34, 35, 36, 37, 38, 39
Read numbers 40 through 49: 40, 41, 42, 43, 44, 45, 46, 47, 48, 49
Read numbers 50 through 59: 50, 51, 52, 53, 54, 55, 56, 57, 58, 59
Read numbers 60 through 69: 60, 61, 62, 63, 64, 65, 66, 67, 68, 69
Read numbers 70 through 79: 70, 71, 72, 73, 74, 75, 76, 77, 78, 79
Read numbers 80 through 89: 80, 81, 82, 83, 84, 85, 86, 87, 88, 89
Read numbers 90 through 99: 90, 91, 92, 93, 94, 95, 96, 97, 98, 99
Rewrite and Memorize numbers from 10 to 99

Three-digit numbers
Three-digit numbers content three digits
This is numbers 100, 101, 102, … 999
Least three-digit number: 100
Greatest three-digit numbers: 999
Digits for three-digit numbers from right to the left:
Ones, tens, and hundreds: 238

Grade 1 Chapter 4 Lesson 10
Summary 4 (Basic Problems)
Comparison numbers

To compare numbers, find what number is greater or smaller or they are equal. One-digit numbers always less than two-digit numbers $9 < 11$

Two-digit numbers always greater than one-digit numbers: $19 > 8$

Compare the tens and then compare the ones. $12 < 18$ $18 > 12$

Addition three numbers in row. Three numbers can add in any order

$9 + 5 + 3 = 14 + 3 = 17$ $9 + 3 + 5 = 9 + 8 = 17$

Subtraction three numbers in row. $10 – 1 – 5 = 9 – 5 = 4$

To subtract numbers, work from left to right. $10 – 1 – 4 = 9 – 4 = 5$

Two operations with three numbers in row

To add and subtract numbers, work from left to right

$5 + 6 – 9 = 11 – 9 = 2$ $5 + 7 – 9 = 12 – 9 = 3$

To subtract and add numbers, work from left to right

$9 - 8 + 7 = 1 + 7 = 8$ $8 - 7 + 6 = 1 + 6 = 7$

Adding tens in column

$$\begin{array}{r}80\\+10\\\hline 90\end{array}\qquad\begin{array}{r}10\\+80\\\hline 90\end{array}$$

Subtracting tens in column

$$\begin{array}{r}90\\-10\\\hline 80\end{array}\qquad\begin{array}{r}90\\-80\\\hline 10\end{array}$$

Adding two-digit numbers in column

Add the ones and then add the tens

$$\begin{array}{r}43\\+35\\\hline 78\end{array}\qquad\begin{array}{r}27\\+52\\\hline 79\end{array}$$

Subtracting two-digit numbers in columns

Subtract the ones then and subtract the tens

$$\begin{array}{r}96\\-25\\\hline 71\end{array}\qquad\begin{array}{r}89\\-52\\\hline 37\end{array}$$

Addition three numbers in column

Add the ones and then add the tens

$$\begin{array}{r}23\\+21\\22\\\hline 66\end{array}\qquad\begin{array}{r}23\\+32\\21\\\hline 76\end{array}$$

Grade 1 Chapter 4 Lesson 10

Table of Addition ((Basic Problems)

$5 + 0 = 5$ $0 + 5 = 5$

Complete the table 1

Number 5 from the left column add number 3 from first row to get answer 8

$5 + 3 = 8$

	+	0	1	2	3	4	5	6	7	8	9
# 1	1										
# 2	2										
# 3	3										
# 4	4										
# 5	5				8						
# 6	6										
# 7	7										
# 8	8										
# 9	9										
# 10	10										

Table of Subtraction

$5 - 0 = 5$ $5 - 5 = 0$

Complete the table 2

$5 + 3 = 8$ $8 - 3 = 5$ $8 - 5 = 3$

	+	?	1	?	3	?	5	?	7	?	9
#1	1									9	
#2	?								9		
#3	3							9			
#4	?						9				
#5	5					9					
#6	?				9						
#7	7			9							
#8	?		9								
#9	9	9									
#10	10										

Grade 1 Chapter 4 Lesson 11
Review 1 A (Basic Problems)
 # 1-20 odd or # 1, 3, 5, 7, 9, 11, 13, 15, 17, 19

 # 1 5 + 0 + 1

 # 2 6 + 1 + 2

 # 3 7 + 2 + 3

 # 4 8 + 3 + 4

 # 5 9 + 4 + 5

 # 6 15 - 0 - 1

 # 7 16 - 1 - 2

 # 8 17 – 2 - 3

 # 9 18 - 3 - 4

 # 10 19 – 4 - 5

Add and Subtract numbers in column

 # 11 $\begin{array}{r} 20 \\ +\ 30 \\ \hline \end{array}$ **# 12** $\begin{array}{r} 42 \\ +\ 25 \\ \hline \end{array}$ **# 13** $\begin{array}{r} 35 \\ +\ 43 \\ \hline \end{array}$ **# 14** $\begin{array}{r} 65 \\ +\ 21 \\ \hline \end{array}$ **# 15** $\begin{array}{r} 34 \\ +\ 43 \\ 21 \\ \hline \end{array}$

 # 16 $\begin{array}{r} 80 \\ -\ 30 \\ \hline \end{array}$ **# 17** $\begin{array}{r} 98 \\ -35 \\ \hline \end{array}$ **# 18** $\begin{array}{r} 48 \\ -\ 23 \\ \hline \end{array}$ **# 19** $\begin{array}{r} 37 \\ -\ 14 \\ \hline \end{array}$ **# 20** $\begin{array}{r} 54 \\ +\ 13 \\ 12 \\ \hline \end{array}$

Grade 1 Chapter 4 Lesson 11
Review 1 B (Basic Problems)
Add and Subtract numbers in row
1-20 odd or # 1, 3, 5, 7, 9, 11, 13, 15, 17, 19

1 5 + 1 + 2

2 6 + 2 + 3

3 7 + 4 + 4

4 8 + 4 + 5

5 9 + 0 + 1

6 15 - 1 - 2

7 16 - 2 - 3

8 17 – 3 - 4

9 18 - 4 - 5

10 19 – 0 - 1

Add and Subtract numbers in column

# 11	40	# 12	53	# 13	35	# 14	25	# 15	14
	+ 20		+ 42		+ 14		+ 32		+ 53
									11
# 16	60	# 17	94	# 18	47	# 19	48	# 20	64
	- 40		-23		-24		- 15		+ 23
									12

Grade 1 Chapter 4 Lesson 12
Final Exam 1 A (Basic Problems)
Add and Subtract numbers in row
 # 1-20 odd or # 1, 3, 5, 7, 9, 11, 13, 15, 17, 19

 # 1 9 + 5 + 4
 # 2 8 + 4 + 3

 # 3 7 + 3 + 2
 # 4 6 + 2 + 1

 # 5 5 + 1 + 0
 # 6 19 - 5 - 4

 # 7 18 - 4 - 3
 # 8 17 – 3 - 2

 # 9 16 – 2 - 1
 # 10 15 – 1 - 0

Add and Subtract numbers in column

# 11 20 + 70	# 12 80 - 50	# 13 15 + 84	# 14 59 - 14	# 15 53 + 32 11
# 16 30 + 50	# 17 80 - 60	# 18 26 + 63	# 19 48 - 15	# 20 42 + 12 23

Grade 1 Chapter 4 Lesson 12
Final Exam 1 B (Basic Problems)

Add and Subtract Numbers in row

1   9 + 1 + 0
2   8 + 1 + 2

3   7 + 2 + 3
4   6 + 3 + 4

5   5 + 4 + 5
6   19 - 1 - 0

7   18 – 2 - 1
8   17 – 3 - 2

9   16 – 4 - 3
10   15 – 5 - 4

Add and Subtract Numbers in column

# 11 40	# 12 60	# 13 37	# 14 47	# 15 24
+ 20	- 10	+ 42	- 12	+ 12
				43
# 16 20	# 17 50	# 18 48	# 19 26	# 20 15
+ 70	- 10	+21	- 14	+ 31
				32

 # 1 - # 20 odd or # 1, 3, 5, 7, 9, 11, 13, 15, 17, 19
Grading Scale: A =10 points, B=9points, C=8 points, D=7 points, F=0-6 points

Final Exam 1 B (Basic Problems)

# 1	10	# 2	11	# 3	12	# 4	13	# 5	14
# 6	18	# 7	15	# 8	12	# 9	9	# 10	6
# 11	60	# 12	50	# 13	79	# 14	35	# 15	79
# 16	90	# 17	40	# 18	69	# 19	12	# 20	78

Grade 1 Answers Chapter 4

Lesson 1

	a	b	c	d		a	b	c	d
# 1	0	1	2	3	# 11	4	5	6	7
# 2	1	1	1	1	# 12	1	1	1	1
# 3	2	3	4	5	# 13	6	7	8	9
# 4	0	1	2	3	# 14	4	5	6	7
# 5	4	4	4	4	# 15	4	4	4	4
# 6	5	6	7	8	# 16	9	10	11	12
# 7	0	1	2	3	# 17	4	5	6	7
# 8	7	7	7	7	# 18	7	7	7	7
# 9	8	9	10	11	# 19	12	13	14	15
# 10	0	1	2	3	# 20	4	5	6	7

Lesson 2

	a	b	c	d		a	b
# 1	8	9	10	11	# 11	0	0
# 2	1	1	1	1	# 12	10	11
# 3	10	11	12	13	# 13	10	11
# 4	8	9	10	11	# 14	3	3
# 5	4	4	4	4	# 15	4	4
# 6	13	14	15	16	# 16	10	11
# 7	8	9	10	11	# 17	10	11
# 8	7	7	7	7	# 18	7	7
# 9	16	17	18	19	# 19	8	8
# 10	8	9	10	8	# 20	9	9

Lesson 3

	a	b	c	d		a	b	c	d
# 1	8	9	10	11	# 11	12	13	14	15
# 2	1	1	1	1	# 12	1	1	1	1
# 3	8	9	10	11	# 13	12	13	14	15
# 4	3	3	3	3	# 14	3	3	3	3
# 5	8	9	10	11	# 15	12	13	14	15
# 6	5	5	5	5	# 16	5	5	5	5
# 7	8	9	10	11	# 17	12	13	14	15
# 8	7	7	7	7	# 18	7	7	7	7
# 9	8	9	10	11	# 19	12	13	14	15
# 10	7	9	9	10	# 20	9	9	9	9

Grade 1 Answers Chapter 4

Lesson 4

	a	b	c	d		a	b	c	d
# 1	16	17	18	19	# 11	0	0	0	0
# 2	1	1	1	1	# 12	16	17	18	19
# 3	16	17	18	19	# 13	2	2	2	2
# 4	3	3	3	3	# 14	16	17	18	19
# 5	16	17	18	19	# 15	4	4	4	4
# 6	5	5	5	5	# 16	16	17	18	19
# 7	16	17	18	19	# 17	6	6	6	6
# 8	7	7	7	7	# 18	16	17	18	19
# 9	16	17	18	19	# 19	8	8	8	8
# 10	9	9	9	9	# 20	16	17	18	19

Lesson 5

	a	b	c	d	e		a	b	c	d	e
# 1	10	5	11	6	0	# 11	15	5	16	11	0
# 2	12	13	5	14	9	# 12	17	18	5	19	14
# 3	9	6	11	5	1	# 13	14	6	16	10	1
# 4	12	12	6	14	8	# 14	17	17	6	19	13
# 5	8	7	11	4	2	# 15	13	7	16	9	2
# 6	12	11	7	14	7	# 16	17	16	7	19	12
# 7	7	8	11	3	3	# 17	12	8	16	8	3
# 8	12	10	8	14	6	# 18	17	15	8	19	11
# 9	6	9	11	2	4	# 19	11	9	16	7	4
# 10	12	9	9	14	5	# 20	17	14	9	19	10

Lesson 6

	a	b	c	d	e		a	b	c	d	e
# 1	10	5	11	11	0	# 11	15	5	16	16	0
# 2	7	13	5	14	14	# 12	12	18	5	19	19
# 3	10	6	10	11	1	# 13	15	6	15	16	1
# 4	6	13	6	13	14	# 14	11	18	6	18	19
# 5	10	7	9	11	2	# 15	15	7	14	16	2
# 6	5	13	7	12	14	# 16	15	7	14	16	2
# 7	10	8	8	11	3	# 17	15	8	13	16	3
# 8	4	13	8	11	14	# 18	9	18	8	16	19
# 9	10	9	7	11	4	# 19	15	9	12	16	4
#10	3	13	9	10	14	# 20	8	18	9	15	19

Lesson 7 Review 4 A Grade 1 Answers Chapter 4

# 1	6	# 2	8	# 3	10	# 4	12	# 5	14
# 6	19	# 7	6	# 8	18	# 9	7	# 10	19
# 11	11	# 12	8	# 13	15	# 14	11	# 15	7
# 16	15	# 17	9	# 18	8	# 19	18	# 20	9

Review 4 B

# 1	7	# 2	6	# 3	5	# 4	4	# 5	3
# 6	18	# 7	6	# 8	16	# 9	5	# 10	17
# 11	4	# 12	7	# 13	13	# 14	7	# 15	7
# 16	12	# 17	7	# 18	6	# 19	16	# 20	10

Lesson 8 Test 4 A

# 1	3	# 2	6	# 3	7	# 4	8	# 5	6
# 6	10	# 7	9	# 8	12	# 9	4	# 10	14
# 11	9	# 12	7	# 13	17	# 14	10	# 15	8
# 16	15	# 17	8	# 18	9	# 19	18	# 20	10

Test 4 B

# 1	2	# 2	7	# 3	8	# 4	9	# 5	5
# 6	10	# 7	8	# 8	12	# 9	5	# 10	14
# 11	7	# 12	8	# 13	17	# 14	9	# 15	10
# 16	15	# 17	7	# 18	8	# 19	18	# 20	11

Lesson 9 Summary 1

	a	b	c	d		a	b	c	d	e
# 1	16	17	18	19	# 6	11	9	16	7	4
# 2	8	9	10	8	# 7	17	14	9	19	10
# 3	16	17	18	19	# 8	9	18	8	16	19
# 4	8	8	8	8	# 9	15	9	12	16	4
# 5	16	17	18	19	# 10	8	18	9	15	19

Lesson 10 Table 1

	+	0	1	2	3	4	5	6	7	8	9
# 1	1	1	2	3	4	5	6	7	8	9	10
# 2	2	2	3	4	5	6	7	8	9	10	11
# 3	3	3	4	5	6	7	8	9	10	11	12
# 4	4	4	5	6	7	8	9	10	11	12	13
# 5	5	5	6	7	8	9	10	11	12	13	14
# 6	6	6	7	8	9	10	11	12	13	14	15
# 7	7	7	8	9	10	11	12	13	14	15	16
# 8	8	8	9	10	11	12	13	14	15	16	17
# 9	9	9	10	11	12	13	14	15	16	17	18
#10	10	10	11	12	13	14	15	16	17	18	19

Lesson 10 Table 2

	+	0	1	2	3	4	5	6	7	8	9
#1	1	1	2	3	4	5	6	7	8	9	10
#2	2	2	3	4	5	6	7	8	9	10	11
#3	3	3	4	5	6	7	8	9	10	11	12
#4	4	4	5	6	7	8	9	10	11	12	13
#5	5	5	6	7	8	9	10	11	12	13	14
#6	6	6	7	8	9	10	11	12	13	14	15
#7	7	7	8	9	10	11	12	13	14	15	16
#8	8	8	9	10	11	12	13	14	15	16	17
#9	9	9	10	11	12	13	14	15	16	17	18
#10	10	10	11	12	13	14	15	16	17	18	19

Lesson 11 Review 1 A (Basic Problems)

# 1	6	# 2	9	# 3	12	# 4	15	# 5	18
# 6	14	# 7	13	# 8	12	# 9	11	# 10	10
# 11	50	# 12	67	# 13	78	# 14	86	# 15	98
# 16	50	# 17	63	# 18	25	# 19	23	# 20	79

Review 1 B (Basic Problems)

# 1	8	# 2	11	# 3	15	# 4	17	# 5	10
# 6	12	# 7	11	# 8	10	# 9	9	# 10	18
# 11	60	# 12	95	# 13	49	# 14	57	# 15	78
# 16	20	# 17	71	# 18	23	# 19	41	# 20	99

Lesson 12 Final Exam 1 A (Basic Problems)

# 1	18	# 2	15	# 3	12	# 4	9	# 5	6
# 6	10	# 7	11	# 8	12	# 9	13	# 10	14
# 11	90	# 12	30	# 13	99	# 14	45	# 15	96
# 16	80	# 17	20	# 18	89	# 19	33	# 20	77

Final Exam 1 B (Basic Problems)

# 1	10	# 2	11	# 3	12	# 4	13	# 5	14
# 6	18	# 7	15	# 8	12	# 9	9	# 10	6
# 11	60	# 12	50	# 13	79	# 14	35	# 15	79
# 16	90	# 17	40	# 18	69	# 19	12	# 20	78

Students!! Attention!!

Adults, help please students 1 – 4 grades to read and understand advice.

When I was in 5th grade I did not like math. I am not always understood math. I was 11 years girl. Next 6th grade I got a very good teacher. She explains very clearly and I understood her very well. I was worked very hard and I was not afraid to ask questions. Final Grade in Algebra in 6th grade I got an A. I said in class that I want to be a teacher of math. But some students said that I am not enough smart to be a math teacher. I said I will be work hard with math. I did it, I got a Master's degree in math and education, I love my students, I was teaching many years, in High School, part-time college math instructor, full-time university math lecturer.

Some people do a simple job, eat, sleep, go to work. People with a bad brain have not so a good job, not so good salary. If you look at TV how someone's dance, you will be not changing your brain. But if you take a dancing class this helps your body and brain ability. The same if you want to use the calculator, but the calculator not always with you. Much better if you learn how you need to do any operations with numbers without a calculator. If you read this math book you will know what you need to do. This teaches you how to improve your memory, attention, logic thinking, how to read your math books every day, your brain works better. You can work 50 minutes with math, 10 minutes break maybe dance. Do not afraid to do mistakes, fixed and move to the next step. Don't listen to anyone, choose what do you like more.

Students need to know if you work with the brain every day, your brain works better. You need a good brain in any job, in any class. When you finish High School, you need to get any profession, you need to go to college and get a good job and a better salary. You need to learn any subject, its work for the brain. Read the book, do the exercises yourself, try your best, and then check your answer at the end of the chapter. Be honest with yourself. You need to believe in yourself. The brain is like muscles. People with good brain lives longer. The professors at my university were over 80 years old, wise, and optimistic. It was a great pleasure to talk to them.

Made in the USA
Monee, IL
05 November 2026

67967404R10057